GW00727967

Two-Way Cable Television

Experiences with Pilot Projects in North America, Japan, and Europe

Proceedings of a Symposium
Held in Munich, April 27–29, 1977

Edited by W. Kaiser, H. Marko, and E. Witte

With 70 Figures and 8 Tables

Springer-Verlag
Berlin Heidelberg New York 1977

Prof. Dr. Wolfgang Kaiser
Institut für Nachrichtenübertragung, Universität Stuttgart
Breitscheidstraße 2, 7000 Stuttgart 1

Prof. Dr. Hans Marko
Lehrstuhl für Nachrichtentechnik, Technische Universität München
Arcisstraße 21, 8000 München 2

Prof. Dr. Eberhard Witte
Institut für Organisation, Universität München
Amalienstraße 73 b, 8000 München 40

ISBN 3-540-08498-3 Springer-Verlag Berlin Heidelberg New York
ISBN 0-387-08498-3 Springer-Verlag New York Heidelberg Berlin

Library of Congress Cataloging in Publication Data. Main entry under title: Two-way cable television. Bibliography: p. Includes index. 1. Community antenna television. I. Kaiser, W. II. Marko, Hans. III. Witte, Eberhard. TK6675.T87 621.388 77-13725

Offsetprinting and Binding: Julius Beltz/Hemsbach
2142/3140-543210

Contents

Introduction

Hans Marko
Munich, Federal Republic of Germany

1. As a charter member of the "MÜNCHNER KREIS" and the organizer of this symposium, I take pleasure in welcoming you to these rooms of the Carl Friedrich von Siemens Foundation.

2. I am pleased that we have succeeded in gathering together such a large and distinguished body of professionally qualified people to discuss new developments and directions in broadband communication.

3. This is the first symposium of the MÜNCHNER KREIS, a supranational organization whose aim is to provide an international forum where not only the technical aspects but also the social and economic implications of new communications media might be discussed on an interdisciplinary basis. Professor Witte, Chairman of the MÜNCHNER KREIS, will have more to say concerning the organization's goals.

4. The physical world in which is live is hallmarked by two major forces: energy and information. From the sociological and economic standpoints, they can equally be viewed as major needs.

5. Whereas in the energy sector there are restrictions and limitations that curb development at every turn, this is not the case in the information sector. On the contrary - information engineering offers a host of new possibilities, many of which are realizable now or in the near future, thanks to new technologies. Among the more obvious examples are semiconductor technology, electronic computers, satellite engineering, and glass fiber transmission systems.

6. Thus while the Club of Rome is compelled to warn of the limits of growth as respects our energy resources, the MÜNCHNER KREIS is confronted with a rash of unique possibilities in the communications area, some of which could well become substitutes for energy consumption. To discern the best possible future and to work toward its realization is the problem.

7. The broadband CATV distribution system with return (upstream) channel is a particularly timely subject in this context. All around the world pilot projects, such as that recommended in the Federal Republic of Germany by the KtK (Commission for the Development of Technical Communication Systems), are being planned or have already been begun. It is our hope that during this symposium we can exchange ideas with the experts working on such projects, and discuss future aspects with them. This discussion will necessarily revolve in large measure around the great unknown - that is, the question of the relationship between technical and economic feasibility and social need or demand. More far-reaching problems, such as legal and political considerations, can and should be given only passing mention.

8. One aspect seems to me deserving of special notice. This is the lack of symmetry in today's mass communication media. These essentially form a system for information distribution only, and therefore provide no opportunity for genuine communication. The consumer's role consists solely of tuning programs in or out, and switching channels. Whether today's passive information consumer will evolve into an at least partially active partner in the communication process is a pivotal question. As expressed in the terms of the MÜNCHNER KREIS: "What shall we do with the return channel?" I hope that the present symposium will give us the answer.

9. The goal of the symposium -- discussion -- demands a drastic limitation of the number of participants. I therefore apologize to all who, despite great interest, are unable to take part. We will try to make the symposium report, which will be published by Springer, available as soon as possible. To reap the full benefit of the discussion, we should like to record the proceedings on tape, and request your permission to do so.

10.   Allow me at this time to thank all who have played a part in making this symposium possible. On behalf of the Lecture Committee I thank all who with their advice and experience helped us to plan the program, particularly Mr. Brownstein of the National Science Foundation in the U.S., who unfortunately cannot be here. I am also especially grateful to Dr. Hochmuth for handling all correspondence. I should further like to thank the communications engineering industry for its backing, and the Carl Friedrich von Siemens Foundation for its hospitality in placing these beautiful rooms at our disposal.

The symposium is herewith opened, and I wish you the best of success.

The Münchner Kreis
A Supranational Association for Communications Research

Eberhard Witte
Munich, Federal Republic of Germany

The MÜNCHNER KREIS was founded in September 1974 on the initiative of leading personages in the fields of science, politics, industry, commerce and the communications media with the wholehearted support of the Bavarian Academy of Science.

The Association's aim is to promote research in questions connected with the development, establishment and operation of technical communications systems and their utilization. Special attention is to be paid thereby to human, social, economic and political problems arising out of the introduction of new communications technologies.

The MÜNCHNER KREIS thus operates on an interdisciplinary level. Whereas special aspects of, for example, telecommunications engineering or of policy in regard to the communications media are the concern of the appropriate scientific bodies, the MÜNCHNER KREIS concentrates on integrating the contributions made by the different scientific disciplines with a view to finding solutions to the comprehensive problems caused by the new communications technologies.

The Association pays special attention to the prerequisites required if innovations in communications technology are to be introduced successfully. Its primary concern is with the problem of gaining people's acceptance for the new services. The question we are confronted with here is inhowfar people are willing to see the traditional direct human communication supplemented, or even in part replaced, by a technically provided long-distance communication (telecommunication). This can of course involve a decisive change in people's relationship to their fellow men and possibly bring about a certain "isolation". On the other hand, telecommunications facilitates "understanding" between people who are separated by distance and would otherwise have no possibility of communicating.

The average citizen will be more and more confronted with the need to select from the mass of information with which he is bombarded from all sides every day. Since people are finding it ever more difficult to cope with the conventional means of documentation and information, one of the tasks of the MÜNCHNER KREIS is to examine the possible new technologies with a view to ascertaining to what extent their implementation would facilitate the more effective utilization of this floor of information.

Besides devoting its attentions to the prerequisites if innovations, the Association also closely studies the effects of new information technologies. It goes without saying that the further development of communication systems will have a decisive impact on economic and social processes within the state, the local community and the family. Serious attention must be paid to the interconnections between communication and democracy.

Since the MÜNCHNER KREIS is a supranational association for communications research, it regards as another of its tasks the discussion - beyond the national borders - of the integration problems outlined above and the consideration of alternative solutions in close collaboration with experts from other countries. The Commission for the Development of the Telecommunication System (KtK) appointed by the Federal Government of Germany had to restrict its work to the telecommunication system within the Federal Republic, but arrived at the conclusion that true telecommunication implies international standardization, on the one hand, and economic and political coordination, on the other.

The recognition is gaining ground that it is imperative to develop a worldwide telecommunications infrastructure which will guarantee the international accessibility of all its members, online terminals and forms of communication which are available to all. The international homogeneity of the legal, organizational and commercial conditions of use should also be insured.

If we are successful in meeting the need for supranational solutions, then telecommunication will itself be in a position to stimulate positive effects which will help assimilate the social and economic structures in the various countries. Telecommunication therefore requires supranational cooperation on the one hand, but also promotes supranational cooperation by creating the possibility of communication and understanding on the other.

The MÜNCHNER KREIS is aware of the fact that it is providing a service. A private association which unites representatives of science, commerce, industry, social policy, the media, law and politics is undoubtedly in a position to send out impulses to promote advances, to elaborate alternative decision possibilities and to forecast their effects. It cannot, however, make the decisions by itself. This is the dividing line between the stimulation offered by expert advisers and the responsibility borne by the appropriate political bodies and institutions.

The aim of the MÜNCHNER KREIS is to provide a forum for the critical discussion of the entire spectrum of well-founded opinion. It will assist the responsible authorities at the national and supranational levels with its solutions and will even press for solutions, but it will hold aloof from partypolitical attachments.

As far as the man in the street is concerned, the MÜNCHNER KREIS wishes to give him a clear, easily comprehensible picture of what the teleco mmunications technologies have in store for the future to enable him to understand possible innovations and the perils they involve. Hence the MÜNCHNER KREIS has a dual goal: that of preparing the general public for the innovative process and that of registering and predicting people's reactions to the new possibilities of communication.

Working Method

The MÜNCHNER KREIS works on the principle of impartiality in politics and of supranationalism in regard to purely national interests and concerns. In this way it secures the freedom to discuss open-mindedly and without considering the interests of other parties any ideas, alternative proposals and problems that may arise and to offer those officially responsible new impulses and promising suggestions.

Due to the supranational and interdisciplinary nature of its tasks, the Association aims at an international membership drawn from the widest possible variety of fields of specialization so that it will be able to deal with the problems of communication on as comprehensive a basis as possible.

The Association's activities concentrate on the following in particular:

1. Symposia are held in which concrete and carefully outlined problems in the field of telecommunication are treated. (Example: the symposium dealing with an exchange of experiences in regard to the planning and implementation of pilot projects of cable communication with return channels.

2. While the symposia are intended as an opportunity for the exchange of ideas in a more limited circle, the Congresses of the MÜNCHNER KREIS are aimed at the general public interested in such problems. (Example: the Congress "Communication and Democracy".)

3. The symposia, congresses and statements of the MÜNCHNER KREIS are prepared and evaluated in study groups (examples: study groups for broadband and narrowband communication). The association also promotes scientific experiments accompanying the field research of new communications technologies and contents.

4. In keeping with its articles of association, the MÜNCHNER KREIS not only organizes events of its own, but also induces external organizations to develop and promote innovations (example: study group "Viewdata").

5. As a result of the scientific analyses and discussions pertaining to communications policy, the MÜNCHNER KREIS submits memoranda, proposals and public statements.

6. The publications of the MÜNCHNER KREIS include Lecture Notes covering the results of symposia and workshops (mainly in English). Over and above this, comprehensive publications on the development of communication policy are directed at a more comprehensive public

Organization

The MÜNCHNER KREIS is a registered association with its office in Munich. Its executive organs are the General Meeting of its members and the Managing Board. The MÜNCHNER KREIS works in the public interest and has been accorded recognition as a non-profit and non-commercial organization.

The work of the Association is financed entirely by the voluntary contributions of its members and by donations. Both natural and juristic persons from Germany and abroad are eligible for membership. New members are admitted by resolution of the General Meeting, their application for membership having been submitted in writing beforehand.

The scientific work of the Association is directed by a Research Committee whose task it is to supervise and conduct the scientific activities of the Association and which, in concurrence with the Managing Board, is authorized to award research projects to third parties. Individuals who are authorities on specific aspects of communication studies and who may also be non-members can be called upon to act as consultants in assisting the Research Committee.

What is technically possible, economically reasonable, socially desirable and politically feasible must be established and critically examined in supranational discussion and by scientific analysis. The MÜNCHNER KREIS is endeavoring to make its contribution to this comprehensive goal.

Der Münchner Kreis
Übernationale Vereinigung für Kommunikationsforschung

Aufgaben

Der MÜNCHNER KREIS wurde im September 1974 auf Initiative von Persönlich-
keiten aus Wissenschaft, Politik, Wirtschaft und den Medien mit Unterstützung
der Bayerischen Akademie der Wissenschaften gegründet.

Der Verein dient der Förderung der wissenschaftlichen Erforschung aller mit
der Entwicklung, der Errichtung und dem Betrieb von technischen Kommunikationssystemen und deren Nutzung zusammenhängenden Fragen. Dabei sollen insbesondere die mit der Einführung neuer Kommunikationstechnologien auftretenden
menschlichen, gesellschaftlichen, wirtschaftlichen und politischen Probleme
behandelt werden.

Die Aufgabenstellung des MÜNCHNER KREISES ist also überdisziplinär. Während
die speziellen Fachaspekte, z. B. der Nachrichtentechnik oder der Medienpolitik,
in den jeweils spezifischen wissenschaftlichen Vereinigungen behandelt werden,
ist der MÜNCHNER KREIS bemüht, die Beiträge der wissenschaftlichen Diszi-
plinen zur Lösung des umfassenden Problems der Kommunikation zu integrieren.

Besonderes Augenmerk wird den Voraussetzungen gewidmet, unter denen Inno-
vationsschritte der Kommunikationstechnologie erfolgreich vollzogen werden
können. Dabei steht das menschliche Akzeptanzproblem im Vordergrund. Es
wird gefragt, inwieweit der Mensch bereit ist, die unmittelbare Humankommu-
nikation durch technisch unterstützte Fernkommunikation (Telekommunikation)
zu ergänzen bzw. teilweise zu ersetzen. Durch diesen Vorgang kann das Ver-
hältnis zum Mitmenschen entscheidend berührt werden und insoweit eine gewisse
"Vereinsamung" entstehen. Andererseits erlaubt die Telekommunikation eine
"Verständigung" von Menschen, die weit voneinander entfernt sind und ohne
technische Unterstützung auf Kommunikation verzichten müßten.

Der Bürger wird in zunehmendem Maße veranlaßt, eine Auswahl aus der Infor-
mationsflut der Massenkommunikation zu treffen, die täglich aus allen Richtungen

auf ihn einwirkt. Mit den konventionellen Techniken von Dokumentation und Information wird diese Auswahl immer problematischer. Es gehört deshalb zu den Aufgaben des MÜNCHNER KREISES, mögliche neue Technologien daraufhin zu prüfen, inwieweit mit ihrem Einsatz das Informationsangebot sinnvoller genutzt werden kann.

Neben den Bedingungen für Innovationsfortschritte werden die Wirkungen der Anwendung neuer Informationstechnologien untersucht. Es ist davon auszugehen, daß von der Weiterentwicklung der technischen Kommunikationssysteme einschneidende Wirkungen auf die wirtschaftlichen und sozialen Prozesse in Staat, Gemeinde und Familie ausgehen. Dem Zusammenhang von Kommunikation und Demokratie ist ernste Aufmerksamkeit zu widmen.

Der MÜNCHNER KREIS als übernationale Vereinigung für Kommunikationsforschung stellt sich die Aufgabe, die Integrationsprobleme der Telekommunikation über die Grenzen hinweg zu diskutieren und mit Experten anderer Länder eng zusammenzuarbeiten. Die von der Bundesregierung eingesetzte Kommission für den Ausbau des technischen Kommunikationssystems (KtK) mußte ihre Arbeit auftragsgemäß auf das Telekommunikationssystem der Bundesrepublik Deutschland konzentrieren, stellte dabei jedoch fest, daß eine wirkliche Fernkommunikation nur durch eine internationale Standardisierung sowie durch ökonomische und politische Abstimmung der Staaten erreicht werden kann.

Es wird zunehmend die Aufgabe erkannt, eine weltweite Infrastruktur der Telekommunikation zu entwickeln, die eine internationale Adressierbarkeit der Teilnehmer, kommunikationsfähige Endgeräte und für jedermann zugängliche Telekommunikationsformen garantiert. Auch die Nutzungsbedingungen in rechtlicher, organisatorischer und wirtschaftlicher Hinsicht sollten international homogen gestaltet werden.

Wenn es gelingt, diese übernationale Aufgabe zu lösen, dann ist die Telekommunikation ihrerseits in der Lage, positive Effekte zur Angleichung der Gesellschafts- und Wirtschaftsordnungen in den einzelnen Ländern auszulösen. Die Telekommunikation bedarf also einerseits der übernationalen Kooperation, andererseits dient sie dieser durch die Schaffung der Möglichkeit zur Verständigung.

Der MÜNCHNER KREIS ist sich bewußt, in diesen Bemühungen eine dienende Funktion zu erfüllen. Eine private Vereinigung, in der Vertreter der Wissenschaft, der Wirtschaft, der Medien, des Rechts und der Politik zusammenwirken, vermag Förderungsimpulse zu vermitteln, Entscheidungsalternativen zu erarbeiten und deren Wirkungen zu prognostizieren. Sie ist jedoch nicht in der Lage, Entscheidungen zu treffen. Hier verläuft die Trennungslinie zwischen fachlicher Anregung und politisch legitimierter Verantwortung.

Der MÜNCHNER KREIS hat den Ehrgeiz, ein Forum der kritischen und vielseitigen Aussprachen zu sein; er wird die Verantwortlichen im nationalen und übernationalen Bereich mit Vorschlägen unterstützen und bedrängen, aber er wird sich der parteipolitischen Stellungnahme enthalten.

Gegenüber dem Bürger wird angestrebt, ihm ein anschauliches Bild über die Zukunft der Telekommunikation zu vermitteln, um die Innovationsmöglichkeiten und Innovationsgefahren plastisch erkennbar werden zu lassen. Damit erfüllt der MÜNCHNER KREIS einerseits die Aufgabe, die Allgemeinheit auf den Prozeß des Fortschritts vorzubereiten; andererseits wird versucht, die Reaktion des Menschen auf neue Kommunikationschancen zu erfassen und vorauszusehen.

Arbeitsweise

Der MÜNCHNER KREIS folgt dem Grundsatz der Überparteilichkeit und der Übernationalität. Dadurch gewinnt er die Freiheit, alle auftretenden Ideen, Alternativen und Probleme aufgeschlossen und ohne Rücksichtnahme zu diskutieren und den offiziell Verantwortlichen neue Impulse und weiterführende Vorschläge zuzuleiten.

Wegen der übernationalen und interdisziplinären Aufgaben bemüht sich die Vereinigung um einen Mitgliederkreis, der die verschiedensten Fachrichtungen repräsentiert, international zusammengesetzt und damit in der Lage ist, das Kommunikationsproblem umfassend zu erörtern.

Im einzelnen konzentriert sich die Arbeit auf die folgenden Aktionen:

1. Es werden Symposien veranstaltet, in denen konkrete, begrenzte Probleme der Telekommunikation bearbeitet werden (Beispiel: Das internationale Symposium zum Austausch von Erfahrungen mit der Planung und Durchführung von Pilotprojekten der Kabelkommunikation mit Rückkanal).

2. Im Gegensatz zu den Symposien, die der internen Diskussion im engeren Kreis gewidmet sind, wenden sich die Kongresse an die fachlich interessierte Öffentlichkeit (Beispiel: Der Kongreß "Kommunikation und Demokratie").

3. In Studiengruppen werden die Symposien, Kongresse und Stellungnahmen des MÜNCHNER KREISES vorbereitet und ausgewertet (Beispiele: Studiengruppen für Breitbandkommunikation und Schmalbandkommunikation). Wissenschaftliche Begleituntersuchungen für Feldversuche mit neuen Kommunikationstechniken und neuen Kommunikationsinhalten werden gefördert.

4. Entsprechend der Satzung des Vereins werden nicht nur eigene Veranstaltungen durchgeführt, sondern auch Impulse zur Bearbeitung und Förderung von Innovationsschritten nach außen gerichtet (Beispiel: Studiengruppe "Bildschirmtext").

5. Als Ergebnisse der wissenschaftlichen und kommunikationspolitischen Analysen und Diskussionen legt der MÜNCHNER KREIS Denkschriften, Vorschläge und öffentliche Erklärungen vor.

6. Die Veröffentlichungen des MÜNCHNER KREISES beziehen sich in Lecture Notes auf die Verhandlungen in Symposien und Arbeitskreisen (vorwiegend in englischer Sprache). Darüber hinaus richten sich umfassende Veröffentlichungen zur kommunikationspolitischen Entwicklung an eine breitere Öffentlichkeit.

Organisation

Der MÜNCHNER KREIS ist ein eingetragener Verein mit Sitz in München. Die Organe sind Mitgliederversammlung und Vorstand. Seine Tätigkeit dient dem Gemeinwohl und ist nicht auf kommerziellen Erfolg ausgerichtet. Der Verein ist als gemeinnützig anerkannt. Die Arbeit des MÜNCHNER KREISES wird ausschließlich durch freiwillige Zuschüsse der Mitglieder und durch Spenden finanziert. Mitglieder des Vereins können natürliche und juristische Personen des In- und Auslandes sein. Die Aufnahme neuer Mitglieder erfolgt durch Beschluß der Mitgliederversammlung; die Aufnahme wird schriftlich beantragt.

Die fachliche Arbeit des MÜNCHNER KREISES liegt in den Händen eines Forschungsausschusses, der die wissenschaftlichen Arbeiten des Vereins betreut und durchführt sowie im Einvernehmen mit dem Vorstand Aufträge an Dritte vergeben kann. Zur Unterstützung des Forschungsausschusses werden Fachbeiräte berufen, die als Einzelpersönlichkeiten für bestimmte fachliche Aspekte der Kommunikation kompetent sind und die auch Nichtmitglieder sein können.

Was technisch möglich, wirtschaftlich sinnvoll, gesellschaftlich wünschenswert und politisch realisierbar ist, muß in übernationalen Diskussionen und fachlichen Analysen ermittelt und kritisch reflektiert werden. Zur Erfüllung dieser umfassenden Aufgaben ist der MÜNCHNER KREIS bemüht, einen Beitrag zu leisten.

Two-Way Cable Television Applied to Non-Entertainment Services: The National Science Foundation Experiments

Charles N. Brownstein
Washington, USA

Introduction

Over the past decade there has been considerable speculation about the potential social benefits of non-entertainment services delivered over cable TV systems. Technological innovation, the growth of the CATV industry, and policies promulgated by Federal authorities raised public expectations, but economic conditions and uncertainty about both market potential and the real utility of new services appeared to retard the development and deployment of new services. This paper describes several active field experiments supported by the National Science Foundation which are designed to generate information about the technical, economic and social feasibility of using commercial CATV systems for public service delivery.

Background

The research discussed in this paper is supported by the National Science Foundation (NSF). Yet the projects are not conducted by NSF, but by independent scholary research scientists. NSF supplies the funds for the projects through its program of Telecommunications Policy Research, which has the mission of supporting research on public service applications of telecommunications technology and on issues of telecommunications policy.

This program of research on applications of cable television was in some ways a response to events and in other ways an attempt to seize an opportunity. The key events leading to the program were: 1) the emergence of very powerful public expectations about what the cable TV industry could potentially provide in terms of beneficial services; 2) the policies of the Federal Communications Commission, particularly those which required new systems to be constructed with bi-directional capabilities; 3) the lack of actual experience (as opposed to speculative studies) with providing services via cable. The major opportunities, which were made possible because of the existence of bi-directional capacity (mostly unused) in existing commercial CATV systems, were: 1) to quickly put together research which could test the efficacy of various telecommunications services, 2) determine empirically the relative costs and benefits of using two-way cable; and 3) to thereby produce information about the social benefits of the FCC policies toward cable.

Given these events and these opportunities, the National Science Foundation issued a call for proposals to design two-way cable experiments. Academic and non-academic research institutions were invited to apply for grants of 6 months duration. Requirements were that any proposal would be on behalf of a consortium consisting of qualified researchers, local government service agencies, and a commercial cable communications company willing to provide the two-way system necessary for the experiment proposed. In this way NSF sought to involve the key actors at the local level in the design of the applications, to insure that they had real commitments to the projects, and to create the essential local infrastructure for the continuation of services which would be discovered to be beneficial during the experiments.

More that 100 proposals to conduct design studies were received. After a very comprehensive review by various experts and by relevant public policy-makers, 8 were selected. The results of the design studies were 8 very comprehensive proposals to implement experiments in actual field settings. Eventually 3 were selected, based on reviewers' judgments about such things: 1) the importance of the applications; 2) the likelihood that the consortium could actually conduct the project; 3) the compatibility of application and technology; 4) the degree to which the research addressed questions of direct relevance to policy-makers; and 5) the degree of community support for the project. Each of these criteria was critical to success. Finally, three projects commenced in the autumn of 1975.

Description of Projects

Each of the projects is a self-contained program of research on particular applications and technologies. However, across all projects, a representative micture of likely technological configurations, of services, and in addition, types of research designs are to be found.

The following is a brief description of each project.

Reading, Pennsylvania: The project in Reading, Pennsylvania tests the use of two-way audio/visual conferencing for providing a variety of social service information programs to the elderly. The technology is point-to-point, with several institutions (residential centers for elderly citizens) interconnected with each other and with the Reading City Hall. Pictures eminating from the nodes of the network are switched or mixed, by special effects such as split screen, manually from the cable head end. Access (origination) is possible from any point, and the resulting picture is sent to selected samples of the subcribers of the commercial cable services. Currently, it is sent to all subscribers (35.000). All programming is live, and much of it consists of people at the various origination points talking to each other.

Some of the programs are focused sharply toward providing information about social services for the elderly, and data are collected on increases in knowledge about the services and on increased use of services. Other programs are more general in nature,consisting of discussions among people at the various points and between the points about local events of interest, and about life experiences. The research design is therefore a mixture of very careful, longitudinal, structured observations of social service agency records, as well as periodic "waves" of interviews of the entire community. The latter are specifically intended as a kind of "net" to capture unanticipated social impacts effecting the larger community as well as the experimental target population. In addition, records of the costs of equipment, man-hours devoted to programming and system operations, and audiences are kept. The effects of the focused information programming is compared with the effects of information provision by conventional means. To determine the costs to the commercial cable television company for use of the cable channel and head end facilities, comprehensive records of incremental costs are also maintained.

Rockford, Illinois: The Rockford project is fundamentally different from the Reading project. A single application, and a more complex technology are tested. The application is in-service training to a single occupational group, firefighters. The training consists of a series of videotaped programs transmitted from the cable system head end. Computer control of the tapes is possible from each of the fire stations, and computer assisted instruction is used as an adjunct to the tapes. Firemen, individually in some cases, and in groups in other cases, view each program and respond to questions by typing in answers. The questions are displayed along with the video tape material. Part of the research concerns this type of technology (computer controlled viewing and computer assisted learning) as it can be made to work on a commercial cable TV system. Part of the research compares the learning process of the mediated and traditional classroom approaches to teaching. Comparisons are also made between individual and group viewing, and between one way only and interactive programs. The application permits a great deal of direct control over participants, and a rather classical experimental design is used. Unlike the Reading project, specially prepared taped program materials are used. These are more costly to produce, but can thereafter be freely available, for use in many settings.

Spartanburg, South Carolina: The project in Spartanburg consists of several experiments testing a variety of applications and technology configurations. One experiment involves formal education to students in their homes. A curriculum consisting of several courses, taught by several teachers, is sent live to student households. Students interact with the instructors by a return channel which accepts digital signals from a compact 8-button terminal. A computer at the head end aggregates the signals, which are displayed in various ways for the instructor. In this way students can respond to multiple-choice questions or they can initiate a variety of communications by sending pre-determined signals. This is compared with conventional classroom instruction.

Other experiments involve multiple-point audio/visual conferencing for training semi-professional workers, and multiple point audio conferencing for teaching various child rearing skills to parents. The conferences are capable of being originated from any point in the cable system. Switching is done both manually from the head end and by remote control from a given point in a conference. Comparisons are made between the teaching effectiveness of audio-only and audio-visual conferences, and between the skills learned by active and passive participants.

Interproject Comparisons

The great differences between the projects were created quite purposely. Given almost complete lack of certainty about the effectiveness of 2-way television for various types of uses, and disagreement over the desirability of various 2-way television technical configurations, it was decided that data on various ranges of key variables should be produced.

For example, one such variable is technology. Across the projects, we have tests of simple and complex digital return path terminals (Spartanburg vs. Rockford), and of multi-point conferencing (audio and video) vs. digital signal polling (in Spartanburg and between Spartanburg and Reading).

Another such variable is type of programming. Live instruction is compared to taped and computer assisted instruction (Spartanburg vs. Rockford). Yet another is the type of participating agency: some are large, others small, some well established, some are invented for the project.

Similarly, a variety of audiences may be contrasted: classrooms for formal education, with informal settings for occupational training (Spartanburg vs. Rockford), and groups with individuals (Rockford).

What will be compared in each case is the overall cost-effectiveness of the two-way over one-way or non-cable means of achieving the goal of communications. These will necessarily be global comparisons and will be suggestive rather than scientifically conlusive. But this part of the analysis will be quite objective - being two steps away from the participating agencies and once removed from the individual project teams and will produce a synthesis of the entire effort. No performer has yet been selected to conduct this part of the program of research. It was unclear when the three projects commenced that it would even be possible to do this type of analysis. But at this time it appears that this synthetic research will be most useful.

Results

The results reported here are necessarily superficial. The experimental portion of the projects has been completed, but data analysis has only just begun. Rather than second guess the researchers, I will offer only a few observations.

First, both the point-to-point and studio-to-many-point applications were technically workable. In Reading and Spartanburg, the "conferencing" applications worked well. The use of a "director" to control special effects and picture switching in Reading appeared to produce more interesting programming than the self switching used in Spartanburg. In both places it was found that it is not necessary that everyone participate for the benefits of 2-way conferencing to occur. Many people participated vicariously, and it was generally agreed that the interactive, live, active nature of the programming made it more involving (even to just watch) than one-way programming. Of course, this represents a very specialized use of a cable TV system as only a small number of nodes can be connected on an interactive basis at one time.

The "general information" programs in Reading have proved to be so useful to the community that the system will be self-financed in the future. The highly focused training programs in Spartanburg do not appear likely to continue. A preliminary conclusion, therefore, is that audience size will continue to be a major factor in new cases of commercial CATV in the U. S. , particularly of 2-way video uses which use a lot of channel capacity.

Second, the "digital return" applications proved technically workable. It appeared that return signalling provided quite adequate interaction between students at home and live teachers at the central studio in Spartanburg. The students were able to direct the teacher's attention to the speed and comprehensability of the instruction, and teachers were able to "collect" responses to multiple-choice questions from both individuals and whole classes. This system is being adopted in Columbus, Ohio, by a commercial CATV system operator. Unfortunately, the results of the Rockford project, which uses more complex terminals and mixes CAI with CATV, are not yet available. The most important observation, however, is that digital return, which can be used simultaneously by thousands of respondents, appears to be "richer" as a communications medium that had been expected.

Implications for Research

A major issue, confronted at the earliest stages of planning for the NSF experiments, was the nature of the research designs required to evaluate the experimental applications. The strategy adopted was based upon a conscious decision to produce research findings which would be useful at several levels: by service delivery agencies, by commercial CATV system owners, by city governments responsible for setting franchise requirements, and for Federal telecommunications decision makers. In particular, it was felt that while research costs incurred in producing the many forms of data needed by these diverse groups of users would be high, that the major cost would be in setting up the operation of the experiments, that is, in designing and delivering the telecommunication service. Thus, it was assumed that the marginal cost of specialized data for any particular need was sufficiently small as to warrant the effort. It appears that this strategy is a good one. Most of the data collection goals have realized, and a wealth of very "fine" data, which are needed to make unequivocal statements about benefits and costs, have been produced. However, not all of these data will be used for each "level" of analysis; different "slices" and aggregations of data will be used for various purposes.

In the case of Spartanburg, early data returns concerning the impact of 2-way video conferencing were used to modify the research plan in order to discover whether audio conferencing (which uses less band-width) would work as well. In Reading, programming was designed by the participants which became quite popular and appears to have produced previously anticipated benefits for the community as a whole. The research design was modified from a true experiment to a quasi-experiment in order to "capture" the effect quantitatively. However, the data concerning productivity, which were produced by the experiment, remain available for the social service agencies which were involved.

The strategy of research used for these projects appears particularly appropriate for the investigation of telecommunication innovations. This strategy effectively maximizes the uses which can be made of the data, and minimizes the changes that unanticipated effects will be captured. In this complex and risky field of research, where research must both test expectations and permit discovery, a mixture of careful scientific design and administrative flexibility is both workable and necessary.

Zusätzliche Dienste beim Zweiweg-Kabelfernsehen – die Feldversuche der National Science Foundation

In den letzten zehn Jahren wurde vielerorts nachgedacht über den möglichen gesellschaftlichen Nutzen von Diensten in Kabelfernsehanlagen, die zusätzlich zu
den üblichen Unterhaltungsprogrammen angeboten werden. Technologische Inno-
vationen, das Wachstum der Kabelfernsehindustrie und die durch die Bundesbe-
hörden getroffenen Festlegungen haben in der Öffentlichkeit Erwartungen geweckt,
jedoch wirtschaftliche Gründe und die Ungewißheit über das Marktpotential sowie
den wirklichen Nutzen derartiger neuer Dienste scheinen deren Entwicklung und
Verbreitung zu hemmen.

Die vorliegende Arbeit beschreibt mehrere Feldversuche auf diesem Gebiet, die
von der National Science Foundation unterstützt werden. Sie sollen Aufschluß
geben über die technische, wirtschaftliche und gesellschaftliche Realisierbarkeit
neuer öffentlicher Dienste in privaten Kabelfernsehanlagen.

A Systematic Plan for Realization of a Full-Service Two-Way Cable System: Four Generations of Technology and Applications

Thomas F. Baldwin
East Lansing/Michigan, USA

This paper first presents a four-generation model for development of two-way cable systems and then describes in detail the first two generations of the plan which are presently operational.

The objective of the model is to provide successive, cost-feasible steps toward achievement of a full-service broadband communication system where independent stages of technical development are matched with service applications. Through the model, we cautiously attempt to work out a set of developmental stages that provide both needed cable services and the potential revenues to accomodate the capital and operating costs of each new stage. The first stage in the model has been operational since 1973 providing per-program pay television services to subscribers in Columbus, Ohio, USA. The second stage has been in operation since February 1977 in Rockford, Illinois, USA, in a computer-managed firefighter training program where the trainees continuously respond to the instructional materials. This project is funded by the federal government's National Science Foundation.

The developmental model

Unfortunately, since the early "blue sky" speculation about broadband communication development, the two-way technology has not evolved as a synthesis of practical need and cost-efficient technology. Two-way experiments and developmental planning have tended to focus on an end-state technology, in particular, a high cost addressable terminal, which in the foreseeable future, is too costly. In the model described below, each generation represents smaller technical steps in the evolution of existing cable technology and service provision. Costs of each step are covered by revenues or benefits accruing from services at that step.

The first generation of our developmental plan includes technology for per-program pay television. The cable industry in the United States is now convincingly demonstrating the demand for pay movies and other entertainment. To achieve the per-program pay television in this generation, we rely on an "area multiplexing" scheme that combines frequency and time division multiplexing. Modified channel converter terminals in the feeder branch are assigned individual frequencies in groups of up to 200. Each group is associated with a digitally controlled code operated switch, which passes signals upstream as the switch is opened. A community of households, each identified by its terminal frequency and code operated switch, can be scanned by a minicomputer in a matter of seconds. If the set is tuned to a pay program, the information is recorded for billing purposes.

The second generation in this developmental model adds an interactive response capability to the terminal. A transmit device permits use of a push button type converter for digital responses as well as channel selection. The additional cost of this capability is low enough to accomodate computer-managed instructional programs which include multiple choice questions. The computer can be programmed to provide a character-generated downstream display of participant responses, question-by-question and cumulatively.

The third generation in the system evolution adds a microprocessor chip and read-only memory (ROM) storage. This permits the terminal to perform multiple functions including the monitoring of utility meters. For electric utilities, the system would offer time-of-day metering, making possible discrimination between energy consumption in peak and off-peak generation periods. Pricing incentives can then be used to encourage consumption in the off-peak periods. This technological generation also accomodates the monitoring of other devices such as smoke detectors and security alarms.

Finally, in the fourth generation, we add low cost memory storage in the terminal or the code operated switch. At this stage it is possible to "page" a portion of a data stream for local, home television display. On-demand catalogs, self-paced questionnaires, on-demand lessons, and electronic newspapers are potential applications in this generation.

At the present time, the first and second generations of the devleopmental model have been achieved and are operational. The third is in the experimental stage. Because we have had operating experience with the first two, they will be discussed more fully.

Per-program pay television

From an economic standpoint, the first generation is a good design for pay television because it permits per-program charges. In April 1977 only two pay cable systems in the United States operated on a per-program basis; one in Allentown, Pennsylvania which began operation early in 1977 and the Columbus, Ohio system which has been operating for four years. A third system, also in Columbus, is scheduled to begin operation in September, 1977. All the other residential pay cable systems provide pay programming on a per-channel basis, where the subscriber pays a monthly fee for access to the channel.

Per-program pay cable is a system of program delivery in which consumers purchase the right to watch each program. This requires two-way technology to monitor the household consumption of pay programs.

The per-channel system does not require two-way capability, only traps or scrambling systems which are less expensive. But, the per-channel pay system is not responsive to viewer demand in a very refined way. On the other hand, per-program pay is sensitive to intensity of program appeal and to price. It provides immediate feedback for use in scheduling (e.g., if demand is high, frequent repeats; if low, few).

The per-program cable system operator is not constrained to making selections and scheduling decisions on the basis of mass audience appeal as is the commercial broadcaster and the per-channel cable operator. For the first time, in United States television, there is a mechanism for reflecting specialized demand for television programs.

Presently, program material (mainly feature films and sports events) is created for other markets. Pay television offers a bonus market. For cable systems, duplication and transmission costs are not large. Syndicators make films available to per-program operators on a percentage-of-gross basis. In a 30- or 35-channel cable system there is little opportunity cost in making several channels available for pay programming. Therefore, the per-program operator has the incentive to offer any kind of programming that will return revenues above the small variable costs.

In summary, the high-capacity per-program pay cable television system solves the economic inefficiencies of the limited-channel commercial television broadcasting industry. By providing a price mechanism to measure consumer program demand, per-program cable increases consumer welfare by encouraging diversity, innovation and minority programming as well as conforming schedules to demand.

The four-year-old per-program pay television system in Columbus, Ohio, bears out some of the projected advantages of the two-way system. Four channels are operating, providing general interest films, cabaret acts, sports events, adult films, "kid stuff" and special interest films. Programming diversity and service of minority interests, is illustrated by programming of the system foreign films as a category of special interest films. While foreign films shown in June 1976 accounted for only .6 percent of total pay television revenues, the per-program system makes it economically feasible to present this narrow interest category of programming.

The system has done some pricing experimentation. We are currently analyzing the pricing and viewer data, by demographic category, to determine elasticity of demand.

Of particular importance to our developmental model, per-program television revenues cover the first generation hardware investment. Hardware costs, primarily expensive terminals in the home, have been a major obstacle to per-program pay television and the reason per-channel pay television predominates. The per-program system used by Telecinema in Columbus, and adapted for the second generation experiments in Rockford, was designed by Broadband Technologies, Inc. The designers sought a low cost terminal. The solution was a combination FDM/TDM "area multiplexing" scheme. It consists of the simultaneous transmission of groups of 200 frequency multiplexed terminals at different time intervals.

Area multiplexing is accomplished through the use of digitally controlled code operated switches (COS). Downstream signals pass through the COS continuously without interruption, while the upstream frequencies are either passed or blocked as directed by a digital signal generated by minicomputer. This allows an entire system of in-home terminals to be scanned in groups of 200 by activating primary COSs in the trunks and secondary COSs at each bridger amplifier.

The COS system permits a much less expensive terminal than a typical interrogation-response terminal because a great deal of circuitry can be eliminated, including the RF receiver, decoder and address-recognition circuitry. A simple FDM terminal requires only a data encoder and RF transmitter circuitry, which can be included in a standard converter. The terminal transmitter is assigned a discrete frequency in its own COS area, and is transmitting all of the time. Each terminal transmits a 16-bit data word which indicates the status of the converter.

A minicomputer manages the entire scanning operation through the use of special interfaces which control a COS addresser and RF receiver. The minicomputer, operating in real time, routinely scans the system collecting data.

Under this system, the converter is modified by adding a circuitry board containing an FSK transmitter and data encoding circuitry costing about $20. A minor additional investment is necessary to take the next step, to the second generation in our developmental model.

Two-way instruction of firefighters

The first generation technology, described above for per-program pay television requires only the monitoring capability to determine the status of the channel converter. A response capability is added with a modification of a push-button type converter. A mode selection switch permits selection of over-the-air television, per-program pay channels or the response mode. A transmit button must be depressed after each converter button response. To insure that the user does not respond more frequently than the minimum scan time, a timed LED display stays on until the scan is complete.

For the second generation, no changes are made in the distribution network of code operated switches. At the headend, more processing is necessary so the minicomputer system must be augmented with additional core memory and sufficient disc storage.

The principal difference between the first and second generations is not in hardware, but computer software. In addition to the basic system scan routing and system maintenance, the computer system must process the response data in real time. We have developed a specialized minicomputer language designed to coordinate two downstream video signals with interactive responses. The software system is capable of accomodating several training programs simultaneously. The computer controls character generation equipment, standard SMPTE time code interfacing with video tape equipment, and standard process control input/output signals and relays to control the necessary video equipment.

The modified converters, used as response terminals are installed in Rockford, Illinois fire stations, to test the system for delivery of instruction. In the experiment, firefighters are being trained in the highly-technical task of prefire planning. The task requires survey of buildings and building perimeters to ascertain construction characteristics, rescue routes, utility locations, location of hazardous materials, etc. To create and read building diagrams, thirty-five symbols must be learned.

The prefire planning training program is a series of 12 videotapes each 35 minutes long. It was determined that prepackaged materials were essential to the training needs of the fire department because (1) there are several working shifts of firefighters located in different stations, (2) there is a continuous influx of new trainees, (3) there is a need for repeated, refresher training, and (4) since firefighters are moved from station to station and often work together in large groups, standardized instruction is necessary. Further, it was decided that, if we could use the two-way system for formative evaluation of instructional programs (pilot testing) with small groups, we could be confident of the pedagogical effectiveness of the material. We would not need the flexibility to make adjustments that live instruction provides.

It is appropriate at this point to describe the lesson procedures in detail. We start a few minutes before the actual lesson with a character generated bulletin which announces the time of the lesson and the fire stations scheduled to participate. There is also some less serious information such as a trivia quiz on fire history, announcements of firefighter birthdays and anniversaries with the fire service. We find that the men are very pleased that the medium can be specifically directed to them, and personalized.

The bulletin information is followed by a countdown tape which starts five minutes before the lesson and allows the firefighters to get settled for the lesson. At the end of the countdown, the men are asked to "log in". Each firefighter has a three letter code which he transmits by the terminal. This in effect, registers the participants. The codes are used later in feedback and record-keeping. As soon as the code is received, it prints on the television screen.

After "log in" the instructional tape starts. For each major instructional point made, about every two minutes, a question is asked. These are usually visual questions with four answer options. The men respond by depressing A, B, C, D, or E (can't decide) on the terminal and then transmit. As soon as a response is received for all codes logged in, the responses of each, by code, appear on the television screens. This confirms receipt of the answers and gives a view of how all others answered. If not all answers have been received in 30 seconds, those responses received to that point are printed. If an individual, or group, fails to respond twice consecutively, they are dropped and the computer thereafter reads only the remaining codes for purposes of advancing the lesson. This is necessary because stations are sometimes called to a fire during a lesson.

After the display of answers, the system returns to videotape, and the narrator discusses the correct answer. The questions have been designed so that they represent small learning steps as in programmed instruction. We designed the programs to have a correct response rate of about 90 percent so that we would have consistent positive reinforcement. After seven lessons we are averaging 88.6 percent, with very few questions falling below the 80 percent level.

At the end of each lesson a "quick quiz" reviews the material presented in the lesson and prior lessons, particularly the prefire planning symbols. These questions are not paced by the participants, but have a fixed five-second period for answering. After the quick quiz, the percent correct for each participant is printed on the screen. The narrator presents the correct answers.

The lesson concludes with a character generated presentation of the percent correct over the whole lesson, combining the quick quiz with all other questions. We also report the percent correct for each shift of firefighters over all lessons to date and the percent correct for each station over all three shifts. Providing this feedback introduces an element of competition which should add interest, if not contribute to learning.

After the lesson has concluded, we provide a few minutes of relaxation for the firefighter by interactive computer games; blackjack, darts, tic-tac-toe, target shooting, etc.

We have developed a language for setting up the lessons in a computer format. The parameters of each lesson are determined and stored--the types of questions, text of character generated messages, color backgrounds, nature of feedback, correct answers, etc. A series of programs constitutes the lesson processor which controls the entire administration of a lesson. The times that specific operations are to be performed are also stored in the computer. During the lesson runs, the separately entered operations and time codes are automatically coordinated.

These operations include starting and pausing the videotape, switching to character generator, scanning the terminals for responses, generating feedback restarting the videotape, aggregating scores and making reports. Once we begin the lesson, everything from the log-in through the final summary report is automatic.

We are testing the viability of the system by using four versions of the training series. There are 53 firefighters in each group. Two treatment groups use the interactive response terminals described earlier to register for the lesson and answer the questions in the videotape. In one of these treatments, a designated person takes the consensus of the viewing group and makes the response to questions on a single terminal. The viewing groups are about four or five persons each. In the other two-way version, each participant has his own terminal. The main point of the comparison between these two groups is to determine whether the people in groups will learn as much as those with individual terminals. We were somewhat fearful that the persons in a group who did not hold individual terminals would become inattentive. We are hoping the two groups do equally well, or even better, that the single terminal groups are superior. Obviously, costs make the individual terminal for each firefighter in a station impractical.

The two other experimental treatments are one-way television. In one version, firefighters respond by paper and pencil to the questions. In the other one-way treatment the firefighters are asked to make a mental note of the answers but are specifically instructed not to make an overt response.

In all four treatments, at the end of a question or question sequence, the narrator discusses the correct answer.

We believe this design to be a severe test of the merit of the two-way system. The quality of the instruction does not vary. All four versions provide the opportunity for continuous interaction with the instructional material. Only the means of response varies.

Learning will be evaluated by gain scores over a 27 item pre- and post-test. A post-test only covers another 49 items. All these items are on prefire planning survey techniques. The test items are presented over television with all four experimental groups responding by paper and pencil. Another set of post-test only items comes from the last lesson in the series which is a review and synthesis of the previous lessons.

Once the firefighters begin making their first prefire plan surveys, by order of the fire chief at the end of the instructional series, a panel of experts will evaluate the survey forms.

Affective instruments are administered before the series, periodically throughout, and at the end. These are designed to assess attitudes and attitude change toward prefire planning, professional firefighting in general, and the mode of training. We are using conventional Likert-type items as well as some of the new ratio scaling techniques developed at Michigan State University. In the latter, people are asked to assign a distance between two concepts on an unbounded numerical scale given a standard.

At this point we do not have a final comparison between the treatment groups in terms of learning or attitudes. We do know, however, that the two-way television is very effective administratively. Most training of firefighters is done by company officers in the station house. The training department must develop the training program and keep track of each company with each of its shifts. In the small city of Rockford, this means administering the training programs for about 40 different units with 40 company officers. Some company officers are effective administrators and instructors and some are not. The two-way television system automates much of the monitoring and record-keeping as well as standardizing the instruction so that the training department knows precisely what has been presented and learned. Immediately after the lesson the computer provides a printout of the performance of each individual, item-by-item and lesson-by-lesson to date in the series. This listing readily indicates the persons who have not completed the lesson, e.g., because of a fire call in the middle of a lesson, sickness, vacation. This is very useful in scheduling make-ups.

We have also learned informally that the firefighters like the two-way training program. They appreciate the challenge, the opportunity for active participation, and the feedback. The element of competition and self-testing, together with the

knowledge that a permanent record of performance is being created, maintains attention, provides a sense of accomplishment, and takes some of the drudgery out of the
very technical instruction.

In calculating costs, if the first generation distribution and headend capital
costs are covered by per-program pay television revenues, the major additional costs
are the response terminals for a few fire stations, the increased computer capacity
and software and the incremental maintenance cost. None of these costs is high and
can easily be accomodated within the fire department training budget, particularly
when the costs are shared by other users such as the in-service training program for
teachers which starts in May 1977.

The production of instructional materials for two-way television is a simple matter. A conventional instructional television format may be employed with the addition of periodic questions at each instructional step. The time code is generated
on the second audio track of the video tape. It would not be difficult to modify
existing instructional materials, such as training films and television programs,
for use on the two-way system. The materials, along with the questions would be
edited onto a three-quarter inch cassette.

The second generation technology in the firefighter training experiment has many
other potentials. In-home applications in the development stages include the testing of preschool children for learning disabilities, school homework assignments
that would be handled much like the firefighter training program, survey research,
and the conducting of public forums.

Third and fourth generation development

Hardware for the third generation two-way development has been designed. Our
most fully developed application is the monitoring of utility meters and alarms.
The energy crisis has heightened interest in time-of-day metering and dual rates to
flatten load curves and make energy generation more efficient. The mushrooming interest in smoke detector alarm systems in the United States provides impetus for
the development of a communication system that links the alarm directly to the fire
department so property is protected, whether or not the residents are present. It
is expected that insurance premium reductions alone could cover the incremental cost
of this service.

It is more difficult to predict implementation of the fourth generation. Technical and production cost breakthroughs must precede it. Nonetheless, we are encouraged by our progress through the first three generations and are convinced of
the value of the small-step developmental model.

Ein systematischer Plan für die Verwirklichung eines alle Dienste umfassenden Zweiweg-Kabelfernsehsystems: Vier Generationen Technik und ihre Anwendungen

Der vorliegende Beitrag stellt zuerst ein Vierstufenmodell zur Entwicklung eines Zweiweg-Kabelfernsehsystems vor und beschreibt dann eingehend die beiden ersten Stufen, die gegenwärtig in Betrieb sind.

Das Ziel des Plans ist die Ausarbeitung aufeinanderfolgender, kostengünstiger Ausbauschritte für ein einsatzfähiges Breitband-Kommunikationssystem, bei dem die einzelnen technischen Entwicklungsstufen den jeweiligen Einsatzbedingungen angepaßt sind. Anhand des Modells wird der Versuch unternommen, eine Serie von Entwicklungsstufen auszuarbeiten, die sowohl die benötigten Kabeldienste als auch die zur Deckung der Kapital- und Betriebskosten jeder neuen Stufe erforderlichen Erträge sicherstellen. Die erste Stufe des Modells, die seit 1973 in Betrieb ist, versorgt die Teilnehmer in Columbus, Ohio, USA, mit Fernsehprogrammen auf Münzbetriebbasis. Die zweite Stufe in Rockford, Illinois, USA, die seit Februar 1977 in Betrieb ist, sendet ein rechnergesteuertes Feuerwehr-Schulungsprogramm, bei dem die Teilnehmer sofort die im Lehrmaterial gestellten Fragen beantworten können. Dieses Projekt wird von der National Science Foundation finanziert.

The Functions of Return Telecommunications for Educational Programming

Judith S. Bazemore and W. A. Lucas
Washington, USA

Discussions of the value of two-way educational programs often leave implicit the assumptions about why two-way telecommunications are necessary. This lack of explicit reasoning makes it easy for those who believe in the value of two-way programs to agree, but it also holds the risk that interactive systems will be designed that serve unnecessary or inappropriate functions for individuals in a particular setting. This paper draws a distinction among three functions of two-way communications for learning, and discusses how they have shaped the educational uses of interactive cable in Spartanburg S. C.

The cable system and community together are the natural laboratory for a series of two-way experiments. The Spartanburg cable system is owned and operated by the TeleCable Corporation which has had considerable experience with two-way cable. Tele-Cable also owns the Overland Park, Kansas cable system which was an early effort with interactive technology, and their experience in Overland Park made a valuable contribution to the design of the Spartanburg system. The technical design of the system relies upon a single cable with low-band split for the return. There is sufficient frequency space for 27 outbound and 4 return video channels. The space of one of the return broadband channels is used for narrowband voice and data communications, and the remaining channels are available for return audio-video broadband communications. The Spartanburg system was turned on in 1972, and the cable now passes in front of approximately 14,000 homes. About half those homes subscribe to cable service.

The operation of two-way educational programs began in Spartanburg in January 1976. All programs have used broadband channels to carry live, instructional classes and workshops, but each application has been designed to test the educational value of different forms of return telecommunications. The two programs we have chosen to treat in this paper involve the simplest and the richest forms of return communications. The adult secondary education program has relied upon a return data capability in student homes, and compared student progress with the performance of students receiving conventional classroom instruction. An inservice training program in child care centers has tested the value of return video transmission in comparison with simple one-way programs without any form of return communications.

The Functions of Return Communications

While there are many ways one might draw conceptual distinctions about the functional uses of return communications, we have found it helpful to distinguish three: program improvement, motivation, and involvement. Different types of return communications have different advantages, and it is instructive to compare data, audio, and video return in terms of these functions. We shall first sketch the relative merits of the alternative forms of return based on our general understanding of the learning process. Then we shall describe how our experience has altered our judgments about their relative value.

Program Improvement. Certainly one major role of student responses is to help the teacher adjust the pace or content of the instruction to the needs of the class. In face-to-face instruction, students barrage the instructor with verbal and non-verbal communications about the state of their understanding of the course content. The instructor can repeat, try different ways of expressing points, give examples, and otherwise tailor the content to the audience. This program function of return communications is distinguished by the fact that the student benefits from the return only if the teacher takes action to alter the course content to meet the student's needs.

When the instructor is communicating over a television channel, the student signals take an electronic form. If the class is at a single location, return video and voice would provide much the same information an instructor would have in the conventional classroom. When one moves to a situation where students are located at several sites, however, return video has only limited value for program feedback because of technical constraints. It is almost impossible to combine effectively video signals from the numerous sites so the teacher must rely on the video return from only two or three sites to judge the appropriateness of the instructional content and style. As a practical matter, return voice is a better source for judging program effects because substantial numbers of students can be given the capability to interrupt and to ask questions about the course content. Since the students cannot see one another, confusion will result from time to time because the students cannot use non-verbal cues to indicate to one another that they are about to speak. But until the class size becomes quite large, voice return can provide sufficient feedback for program guidance (Holloway et al., 1975).

Return data is presently the only way of obtaining direct and simultaneous reactions from very large classes of students. Instructors can make requests or students can on their own initiative send prearranged codes to indicate confusion or to ask simple questions. With the appropriate software, the instructors can be given undated summary statistics or codes attached to specific names. In this regard, return data can provide diagnostics about the performance of the entire class that are superior to the cues available in return voice or even in face-to-face instruction of large classes. The instructor teaching in the classroom or receiving a voice return must rely on cues provided by the more outspoken and actively verbal students. Return data cannot match the ability of voice communications to ascertain the nature of a misunderstanding, but it can provide the capability to make the program responsive to the needs of all instead of a few.

Motivation. A central justification for a return capacity is the role it plays in motivating students to learn. Even highly motivated students must often be

encouraged to deal with difficult materials. In programs for students without a
commitment to education, the instructor's ability to motivate the student may be the
major factor in program success. In absence of any form of return communications,
the televised instructor can express generalized concern about student progress, and
one characteristic of extraordinary teacher-actors is their ability to project empathy
and social support for their students. Nevertheless, so long as the student cannot
transmit a return message, the student knows the teacher cannot be responding to him
as a unique individual.

Return video clearly provides the greatest opportunity for motivation, with audio
not far behind. Return video gives the student confidence that the teacher is responding to the full richness of his requests, and the teacher has the opportunity to respond to non-verbal behavior. Again, for large numbers of students, return voice is
probably more practical. The teacher can switch video from student to student, but
more spantaneity results in voice communications. Even for person-to-person counseling,
analogous to student-teacher discussions before and after class, two-way video may not
offer that much advantage over video with return audio for motivating student achievement.

This function appeared initially to be a fundamental--if not fatal--weakness of
data signal return. A student limited to an alphanumeric keyboard can type and send
extended questions, but that information would not have permitted the rich nuances
of meaning provided by video or voice communications. If the student has only a simple
numeric capability, limited to signaling responses to structured questions and a few
pre-coded messages, the return information flow is highly restricted. It would be
very difficult then to establish the affective links between teacher and student that
are essential to the provision of social encouragement and support.

Experience proved this expectation, more than any other, to be substantially
incorrect. It is important to recognize that the advantage of video and audio return
are realized only by students willing to display interest, give answers, and otherwise participate in the class. Without those behaviors, regardless of the nature of
the return, there is nothing for the teacher to reinforce. By requiring and structuring the participation of usually passive students, return data systems provide a
base of interactive behaviors the teacher can use to motivate students.

Involvement. The third function of the return is the most complex, and it may
be that the label "involvement" does not convey the sense of what we intend. But,
however the concept is defined, involvement appears to be made manifest by at least
two operations which are related to the learning process: attention and imitation.
Following the definition offered by B. R. Bugelski, attention is a result of the control of the student by the operation of dominant stimuli, during which the stimuli
involved can interact, affect each other, or be associated with each other in terms
of their neural activities (1971, p.168). Further, the ability to attend and concentrate, as suggested by Albert J. Harris, means that the student can maintain focus
on particular stimuli and disregard or suppress other stimulation that reaches him
at the same time (1970, p.224).

Attention, as this definition suggests, has implications for both teacher and student. The teacher must know or discover the stimuli that can be manipulated to create a state of attention; the student must be able to maintain the state and respond actively within it. During the learning period, the student must be active or responsive to the material involved. This activity is not necessarily immediately observable, but it must occur and must generate the potential for and interest in future related activity. Students who do not attend, due either to the ineffectiveness of the stimuli or their own inability to maintain focus, are neither active or responsive to the material. Consequently, learning is not likely to occur.

Response to the learning situation often takes the form of imitation, probably the greatest single and most widely practiced operation in learning. If they have been attentive, students can be expected to imitate some behavior. The successful teacher will introduce and make use of specific stimuli to encourage the student's imitative behavior. When these same stimuli occur subsequently in the absence of the teacher, the learned behavior can be expected from the student.

By use of return video with voice and the verbal and non-verbal cues it supplies, the teacher is made more sensitive to the stimuli operating on the individual. Thus the teacher can manipulate the learning environment or subject matter to maximize student attention and subsequent response, provided that the student is not unduly distracted by the additional stimulus of being singled out by the camera and consequently unable to maintain his state of attention. Return video with voice also allows the teacher to monitor the student's overt imitative behavior and provide reinforcement for that behavior, again subject to the student's ability to attend and respond to the situation.

With use of only return voice, the student cannot be distracted by visual images of other students, but the sources of visual cues for the teacher are removed. While this kind of voice feedback from the student suggests his attention and active response to the material, one can only assume satisfactory imitation/performance will occur as a result of his reported attention or involvement. Moreover, since the students cannot see each other and use non-verbal cues to indicate that they are about to speak, the resulting confusion may lessen or eliminate the less extroverted or determined student's attempts to respond. Should this occur, the teacher will not be able to use voice participation as a tool to help to maintain the individual student's attention.

Return data signal enables the student to be actively responsive to the learning situation in ways that help both the student and the teacher. The student can, without threat of exposure to his peers, answer questions and make requests simply by stroking a keyboard. The display of individual student responses helps the teacher by enabling her to concentrate on individual performance without extraneous information.

As with the other two return modes, return data signal has an effect on the amount and kind of imitative behavior which can be observed. Each mode imposes different sorts of limits on both amount and kind of observable imitative behavior. In every case, only immediate imitative behavior can be observed. Observation of delayed imitation, which is a frequent condition of the learning process, is beyond the power of the technology, just as it is often beyond the power of the classroom teacher. One assumes that imitation is the result of attention and is the manifestation of learning. This learning may be purely cognitive style; or it may be both cognitive and social and the imitative behavior be as mundane as duplication of a physical movement or as complex as an imitation and subsequent adoption of a style of social interaction.

Of the three return modes, only return video allows direct observation of the development of behaviors such as psychomotor skills, interactive styles, and cognition as it is displayed in recitation. Having only return voice forces reliance on self-report for all behaviors. Return data signal, again, is highly constrained but is an individual and nonthreatening self-reporting system. When the return video is seen by both the students and teacher, it is a rich source of cues for the teacher but a distraction which might interfere with the student's attention and, thus, his learning or amount of imitative behavior. Although each return mode makes a contribution to program improvement, involvement and motivation, the advantages of each mode appear to be relative to the purpose--what is to be taught and to whom--for which the technology is employed. The question then becomes: Under given circumstances, how much return capacity is needed to accomplish a specified set of learning objectives?

With funding, from the Research Applied to National Needs (RANN) program of the National Science Foundation, Rand began a series of field experiments to determine, among other things, the validity of these arguments. Two such experiments will be reported upon here. Adult education was selected to test the value of return data transmission. Inservice training for the staff of child care centers was used to explore the dynamics of learning with return video.

CHILD CARE INSERVICE TRAINING

In Spartanburg, child care often means little more than a custodial service because the caregiver is either not convinced of the need for educationally rich programs or not trained in how to provide such programs. Limited opportunities for staff training exist in Spartanburg despite the size of the day care community. There are 2,388 children cared for outside the home in 51 licensed facilities in the country. And while there is no accurate count of unlicensed facilities or the children in them, the number of unlicensed facilities adds substantially to these numbers. Although research has indicated that the quality of the caregiving situation is determined in large part by the quality of the staff, licensing regulations are unenforceable and include no more than custodial guidelines. The solution to the problem of upgrading child care services appeared to be the provision of training for caregivers in the areas of program development, adult-child interaction styles, and center management.

The training needs of the Spartanburg day care community, and the decisions about the curriculum content and teaching approach, led in turn to the use of return video and audio, broadband communications. The purpose of the child care inservice training was to facilitate caregiver transactions with children on both the cognitive and the affective levels. Transactions between caregiver and child in a child care center occur in a variety of circumstances and are largely influenced by what the caregiver brings to the transaction. Typically, the caregiver performs one or more of three roles: custodian, monitor, or teacher in the formal sense. Each of these roles requires, on the one hand, a specific degree of cognitive information; on the other hand, all demand effective interpersonal transaction skills if this cognitive information is to be translated into successful practice.

The curriculum should, we felt, deal with both concerns. The overall thrust of the workshops was to provide an interactive learning experience involving both cognition and affect for the caregiver in much the same manner that such an experience can be provided for the child. The curriculum content was to be based on assumptions concerning principles of child development and the characteristics of a desirable child care center environment which are conducive to implementing these principles. The content focused on cognitive and socio-emotional development, and the approach emphasized environmental interactions and quantity and variety of interactions that maximize feedback to the child whether by people or objects in the activity. The parts of the curriculum addressed to the socio-emotional development emphasized chances for the child to enter a variety of social relationships, chances for the child to express his emotions, and receive emotionally honest feedback that included situationally-based reasons, rather than assertions of authority. The curriculum assumed that certain characteristics of child care centers were conducive to implementing these principles of child development. The use of space; health and safety, including nutrition; and staff characteristics were addressed as they related to and were supportive of the principles of child development under study.

The philosophical orientation of the workshops was that the principle and its concrete application should be taught together. Since our approach was that of demonstration and interaction in a workshop setting, the presentation of each curriculum module was organized in three steps: (1) presentation and illustration of principles; (2) demonstration of relationship between principles and performance; and (3) interactive experience in combining principles and performance. For example, the ways a child's development limit his ability to perform was first presented, followed by a demonstration of how to teach a skill to children of different ages, then the caregivers constructed materials for play activities appropriate to different stages of development. The caregiver was thus creating in the form of books, games, puzzles, etc., the means by which she would put the principle into practice for the child. Two-way audio return enabled the person conducting the workshop, as well as the participants from other centers to give the caregiver feedback about his or her performance.

This approach was considered to be particularly valuable in demonstrating the relationship between principles of socio-emotional development and daily transactions with children in her role as either custodian, monitor, or teacher. By means of the two-way return, the participants and the professional could interact on a wholistic basis--receiving a sense of confidence from positive evaluation of work well done, encouragement to express feelings of frustration or dislike, or feedback accompanied by situationally based reasons--in the same manner one should interact with a child. Thus, we modeled on a highly spontaneous basis the kinds of interpersonal transactions that are effective in dealing with children and peers.

The curriculum and teaching philosophy thus placed a heavy emphasis on modeling and imitation of complex social behavior. Return video seemed essential to that purpose, and a return broadband capability was provided to selected day care centers.

Experimental Results

The research design to evaluate this multipoint, broadband return used random assignment to create two treatment groups and a quasi-experimental control group. All child care facilities in the Spartanburg cable area were invited to participate and sixteen agreed to join the program. These centers were put into matched pairs based on several institutional characteristics, and the first or second center in each pair was randomly assigned to the two-way condition. These 8 centers received cameras and modulators, and had the capacity to originate picture and voice signals on a return broadband channel. The remaining 8 were assigned to a one-way condition, and never received any return capacity. In addition, 6 child care facilities outside the cable area were also recruited to join the program. Their personnel constituted a control group for the experiment.

We then asked the directors of the facilities to request the participation of their staffs. The expectation was that "one or two" of each teaching staff would join the program. When the program began, 50 directors and teachers had agreed to be subjects for the study.

All subjects were given a battery of tests. In addition to gathering background data on the subjects, their professional experience and the characteristics of their centers, we administered a situation-based test of cognitive information and we collected four hours of observation data. The information test was essentially a multiple-choice test, but because of the wide variation in eduational backgrounds of the subjects, it took the form of a booklet of cartoons. Each cartoon story created a situation which required a response from a fictitious child care teacher, and the subjects were asked to choose the best response. While the instrument sought to measure several areas of knowledge, over half of the stories concentrated on knowledge of methods and principles for teaching pre-school children.

The observation instruments sought to capture the frequency and quality of interactions between the caregivers and the children. A trained observer recorded every interaction the teacher had, whether it was with a group or with an individual child, and described each interaction in terms of positive or negative affect and communication type. Every five minutes, the observer also noted the nature of the activities, such as art or music, in which the children were engaged and what proportion of the children under the teacher's care were actively involved in these activities. Each teacher was observed on two days, two hours each day, by different observers.

In mid-January, the programming began. Each weekday, Monday through Friday, the program was on the cable from one to two o'clock. A professional led the training activities from either the studio or from one of the centers. The centers with cameras were called on to establish their presence in the day's program and then from time to time to share experiences or questions, or to show whatever teaching devices they might be constructing under the direction of the leader, using the supplies they had been given earlier. All the viewers were encouraged to participate in the activities along with the leader but only those with cameras, the two-way group could be observed participating.

The results confirm the difficulty of using return video and voice effectively. Using the cartoon test of knowledge of teaching concepts and methods, we found that there were experimental effects. Both those with and without cameras who had watched the cable workshops made significant gains above the improvement of the staff in the control group of centers outside the cable area. There were no differences between the two experimental groups, however, for those without cameras learned as much as those with cameras who had participated with return video and voice communications. The fact that the center staff were usually watching in small groups of two or three dictates caution in interpreting the results. The available research has generally shown that in cognitive learning where is no difference in the classroom between a live and a televised teacher (Jamison et al., 1974). In this case, the peer group setting permitted interpersonal interaction which may have served some of the involving and motiviating functions of learning. Since the professionals directing the workshops could get video and voice feedback from some of the centers, that was apparently sufficient to enable them to keep the program content responsive to the viewers' needs.

What was more surprising was the results from the behavior observation instruments. If we had expected a competitive advantage of the return video in this experiment, it was in the area of behavior change. Our curriculum placed a heavy emphasis on modeling behavior, and the explicit use of imitation to try to induce behavioral change. Participants with cameras followed the actions of the workshop leader, imitating what the leader had done, whether it be making and using equipment for a day care activity or interacting with another participant in a particular way. The leader could watch and correct the imitative behavior, and, we believed equally important, the participant would be acting out the behavior in a public way, building external social support for the behavior. Those in the centers without cameras not only were able to view the interaction between the leader and participants in the other centers but also were encouraged to complete exercises and materials along with the workshop leader and those participants with cameras. Nevertheless, there was no evidence of net change in the behavior of either those with or without cameras. They continued to interact with children in much the same pattern as they did before the cable workshops.

Perhaps the presumed advantage of direct interaction with the workshop leader by means of return video was offset by the structure imposed by the presence of the camera. It could be argued that the participants had been forced into a performer role in a peer situation with the professional, instead of being child care workers interacting spontaneously with other caregivers. The direct interaction was also impaired by technical difficulties, occasioned by rotating cameras, manual switching and other features of the system. These distractions and the general obstrusiveness of the cameras may have undercut the possible effects of the programs on behavior patterns. But whatever the force of these explanations for the absence of behavior change, it may be that the only safe conclusion is that day-to-day human interactive behavior is very difficult to alter.

DATA RETURN AND ADULT EDUCATION

Adult education was chosen as the service area for the use of return data systems because it met several objectives of the project. We wished to gain experience with home rather than center-based delivery of services and there was evidence of a market for adult education in the home.

Over half of the adults over 25 years of age in Spartanburg had not completed their secondary education, but very few would come forward to enroll in programs using conventional classroom education. The staff of these programs felt that lack of transportation and the need for child care tied many potential students to their homes. They also reported a general reluctance of adults to admit they had not completed their education, and it was hoped that the relative anonymity of taking a course at home would side-step many of the psychological barriers facing these adults.

After a review of the different approaches to adult education, we choose that of the General Educational Development (GED) offered by Spartanburg Technical College. In the GED program, adults enroll in an intensive, 12 hours a week, 15 week course of instruction in language, reading, and mathematics to prepare themselves to take a GED examination administered by the state educational agency. Those who pass the test are given a GED degree generally accepted as being equivalent to a secondary school diploma. Spartanburg Tec was conducting conventional classes using teaching methods that seemed appropriate for cable television, and they relied heavily upon texts and work materials that included a large number of multiple choice exercises. The nature of those materials, coupled with the fact that the GED test the students were preparing to take is multiple choice, led us to the conclusion that GED education was ideally suited to home education using data terminals for return.

The purpose of the GED classes was to enable the students to perform successfully on the GED test. The curriculum was used as a vehicle for shaping cognitive style and developing a test-taking set, an approach facilitated by the use of return data signal. The curriculum, then, was defined as the total cable classroom experience, which included format, subject matter content, and teaching approach. We will discuss each of these as they relate to the development of cognition.

As mentioned earlier, the format of the GEd curriculum as it was being taught at Spartanburg Tec, appeared to be ideally suited to home education using data terminals for return. Time in the conventional classroom is devoted to many activities including substantive presentation, classroom discussion, individual work periods, drill, procedural information, and examinations. Of these, only the classroom discussion was seen to be technically difficult for the cable classroom. On the other hand, this component was foreign to the actual test situation, and consequently, irrelevant to, perhaps even unfavorable to, a model of the test environment.

The texts, work materials, and the GED test itself require specific answers to objective questions in a multiple choice format. It was not the purpose of the class to teach the answers to the test but rather to teach a way to attack the questions and respond to them. By using these materials to structure the stimulus environment and the learner's reactions to the stimuli, the teacher could assist the student in developing a mental set for test taking. Such a set would result in the student's being able to organize incoming sensory data and respond appropriately to it. In this light, learning is the increasing of the strength of a specific response in a hierarchy of responses and can be monitored by a count of correct responses. For the cable classroom, the computer would record, display, and store responses that provided excellent diagnostics to help the teacher direct student activities.

Organization of incoming sensory data is facilitated by the learner's ability to attend. Until he learns to select from among the stimuli which constantly bombard him, it is well that he be assisted in selection by a reduction in stimuli and by experiencing the stimuli as part of a pattern. Since the GED material can be presented as somewhat narrow specialized bits of information and since it requires one primary response pattern, the number of potential distractions is sharply reduced--a situation which is enhanced by the cable student's being outside the conventional classroom situation and its attendant distractions, both social and intellectual.

The effect of such distractions is, of course, relative to the characteristics of the learner and the effect of his past experiences. Generally, the adult student has had no experience with formal education for at least ten years; if he is employed, it is in a low-salaried service position; his social milieu is sharply defined; and he is defensive about his lack of a diploma, which is often regarded as the symbol of intelligence rather than the certification of school attendance and achievement. Frequently, there is a residue of ill feeling toward school and teachers and that previous educational experience which, for one reason or another, resulted in failure.

A return to the conventional classroom for this adult is a risk. Forgetting the concerns of transporation, child care, appearance, and the admission to one's family and friends that one is not a high school graduate, there are still distractions such as the threat of exposure if one does not answer a question correctly, or the ever-present competition for the teacher's attention.

The cable classroom format does not contain these distractions. In relative anonymity, the cable student can respond to every question, be immediately reinforced for his efforts, and command the teacher's attention for himself alone simply by stroking his keyboard. He selects from among the data he receives, practices the organization of these data, responds to the situation, and using the feedback he receives, is assisted in constructing a cognitive set which will gradually develop activity of its own, needing only the immediate sensory stimulation of subject matter content and test environment to trigger the activity.

The function of the subject matter content of the curriculum is to provide relevant practice material for the learning experience. These materials review the basic concepts and rules of math, language arts, and reading. They introduce the specialized vocabulary of the field and provide the student with experience in discerning various author's styles and reacting to them. Students engage in simulation of the test proceedings using this information as background resource, manipulating the concepts and vocabulary in increasingly more appropriate organizational and response schemes.

Just as the student's behavior is controlled, to an extent, by the return data signal, so is the teacher's approach, the third component of the total cable classroom experience. On the one hand, the teacher must structure and refine his presentation such that the return data signal becomes an effective communication tool. On the other hand, properly used, return data allows the teacher to manipulate stimuli more precisely for each student as it provides instantaneous diagnosis and cues to a student's progress without the distraction of a conventional classroom interaction. Equally vital, the teacher is able to establish affective links with the individual student and reinforce his responses to the same degree every other student is reinforced; with the keyboard as equalizer, the student does not need to compete for the teacher's attention and can devote all of his energy to the learning task.

The research design took advantage both of the short, intensive nature of the course and the fact that it employs three different teachers. On any given day, two different teachers offer, for example, first math and then reading. A schedule could be established rotating teachers between a conventional classroom setting and the cable studio so that both groups of adults used the same texts and received the same instruction. Thus the math teacher would instruct the cable class from 8:30 to 10:00 a.m., and then drive to her school during a break so that she could teach a conventional class from 10:30 to noon. The reading teacher would likewise cover the same material twice, once for the conventional class from 8:30 to 10:00 a.m. and a second time in the studio for the cable class.

In February 1976, the first matched classes were offered. After extensive advertising, a conventional class of 25 and a cable class of 10 adults were enrolled. Two subsequent classes have been offered in the fall of 1976 and the spring of 1977.

Based on the findings of standardized achievement tests, the cable program with return data proved to be as successful as conventional classroom education. The students all took the Adult Basic Learning Examination (ABLE) before their classes began and then were reexamined at the conclusion of the course. We also have their scores on the state GED examination, and we know whether they were successful in obtaining their secondary school equivalency certificate. The cable classes matched the performance of adult students taking the same course from the same teachers in a classroom setting.

CONCLUSION

Telecommunications, like any other form of human interaction, is shaped by the culture and institutions of their society. Spartanburg is a city of 45,000 in the southern United States, and the experience of the two-way cable project being conducted there may or may not be relevant to Europe. It is also important to note that when a scholar thinks of what education ought to be, he or she often thinks in terms of a university seminar where there is a lively exchange of ideas. Whether this image holds true as much as many would like to believe is an open, empirical question, but it is not characteristic of the types of education and training that have been our concern in Spartanburg.

With these caveats in mind, we have found that return data communications are a remarkably powerful educational tool. Large classes can be taught, and the teacher can use responses to alter the content and pacing of the course so that it helps the entire class. Because all the students respond, the teacher has student behavior that can be reinforced to motivate learning. Since all students can and are required to respond to questions, the use of data terminals appears to strengthen and structure attention. Even in the area of imitation and modeling, return data is quite appropriate when the purpose is to present information that will be tested in a structured way.

The only clear advantage of return video with voice appears to be when complex social behavior must be imitated. In this project, however, we have found no evidence that return video is essential. A questionnaire administered to those with cameras at the end of the program found several objections to being on display, suggesting that for some return video was actually worse than having only return audio. And when the participants were asked if they felt that they would have needed a camera or whether a microphone would have been sufficient, they answered overwhelmingly that return audio is enough. Similarly, those without cameras responded that they would strongly prefer to have a return capacity, but when pressed, they too tended to say a voice return would have been sufficient. The conclusion appears to be that the simpler, narrowband forms of return communication are adequate for many educational applications.

Bibliography

1. Bugelski, B.R.: The Psychology of Learning Applied to Teaching, 2nd ed.
   New York: Bobbs-Merrill Company, Inc., 1971 '''

2. Harris, A.J.: How to Increase Reading Ability, 5th ed. New York: David
   McKay Company, 1970 '''

3. Holloway, S., Hammond, S.: Tutoring by Telephone: A Case Study in the Open
   University, Communications Studies Group, Joint Unit for Planning Research,
   P/75025/HL, 1975 '''

4. Jamison, D., Suppes, P., Wells, S.: The Effectiveness of Alternative Instructional Media: A Survey,. Review of Educational Research. 44, (1974)'

Die Funktion der Zweiweg-Kommunikationssysteme für Bildungsprogramme

Diese Arbeit berichtet über die in Spartanburg gewonnenen Erfahrungen mit einem Zweiweg-Kabelfernsehsystem, bei dem die Lernenden durch ein Dialogsystem aktiv in das ablaufende Lehrprogramm eingreifen konnten. Die relativen Vorteile der doppelt gerichteten Daten-, Bild- und Tonübertragung werden in Bezug auf die drei erzieherischen Funktionen - Programmverbesserung, Motivierung und Beteiligung der Lernenden - untersucht. Zwei der in Spartanburg verwendeten Dialogprogramme werden anschließend unter diesem funktionellen Aspekt untersucht.

Programme für die Erwachsenenbildung sind abhängig von Datenendgeräten beim Teilnehmer. Dadurch ist es den Lehrern möglich, die Programminhalte anzupassen und eine hohe Beteiligungsrate bei den Studenten zu erreichen. Die Endgeräte stellen auch eine Hilfe für die Lehrer hinsichtlich der Motivierung der Lernenden dar. Im allgemeinen hat sich der Unterricht mit Datendialogsystemen als ebenso wirksam wie der konventionelle Unterricht erwiesen.

Beim praktisch orientierten Ausbildungsprogramm wurden Zweiweg-Bild- und Tonsysteme für die Kommunikation zwischen Kinderbetreuungsstätten verwendet. Das System war anfänglich umständlich zu bedienen und arbeitete nicht immer zuverlässig, so daß Schlußfolgerungen zwangsweise als provisorisch zu betrachten sind. Die größte Stärke des Zweiweg-Fernsehsystems scheint beim Einsatz als billiges Mittel zur Herstellung von Programmen zu liegen. Beim Ausbildungsprogramm für die Kinderbetreuung wurde Teilnehmern ohne Kamera, welche den Dialogverkehr der anderen Teilnehmer über Kabelfernsehen beobachteten, ebensoviel Wissen vermittelt wie jenen mit Kamera. Man war bisher der Meinung, daß sich Fernseh-, Bild- und Tonsysteme vor allem dort vorteilhaft einsetzen lassen, wo kompliziertes soziales Verhalten imitiert werden muß. Es ergaben sich jedoch keine Anhaltspunkte für die Notwendigkeit eines Zweiweg-Fernsehsystems. Es scheint der Schluß nahezuliegen, daß die einfacheren Zweiweg-Kommunikationssysteme im Schmalbandbereich für viele Anwendungsmöglichkeiten im Bildungs- und Erziehungsbereich ausreichend sind.

Interactive Telecommunications and Local Community Processes

Mitchell L. Moss
New York, USA

Two-way cable television has vast potential for addressing the communication needs of both the private and public sectors in urban areas. Although there has been extensive discussion of potential applications of two-way cable television, few public uses have actually been developed. This can largely be attributed to uncertainty about the costs and benefits of such applications. This paper reports on an experiment which is supported by the National Science Foundation and designed to assess the role of interactive telecommunications in the provision of urban services. The project is being conducted by the New York University Alternate Media Center, School of the Arts and the Graduate School of Public Administration over a thirty-month period.

This experimental system is a prototype for public uses of two-way cable television in urban communities. The purpose of the research is twofold: to determine the costs and benefits of using interactive telecommunications to deliver public services and to evaluate the impact of this communication technology on citizens and urban service delivery organizations.

New York University, in collaboration with local public and private organizations, has designed and installed an interactive cable television system. The system has been established to serve senior citizens and public agencies in Reading, Pennsylvania, an industrial city of 88,000 located approximately 60 miles northwest of Philadelphia. The

system consists of three interconnected neighborhood communication centers which are located in a multi-service center and two senior citizen
housing projects. Local government offices and high schools are also
connected to the system as are the homes of local cable-subscribers.
Programming over the system is designed to provide information and
referral for social services, education and training, and citizen-
government interaction.

The neighborhood communication centers are equipped with portable television cameras and monitors which permit two-way communication among
the three centers. Initially, special converters were installed in the
private homes of approximately 125 elderly citizens to allow them to
view the cable programming over their home television sets and to participate by telephone. The positive response by homeviewers to the
interactive programming led to the subsequent decision to extend the
programming to the 35,000 local cable television subscribers.

Programming over the system consists of daily, interactive sessions
which originate from the neighborhood communication centers as well as
from various remote locations such as the City Hall, the local office
of the Social Security Administration, the County Court House, and several high schools. The programs, which are transmitted two hours a day,
five days a week, include citizen-government interaction, information
and referral on social services, and education and training.

The basic methodology for assessing the effect of the two-way cable
television system employs treatment and control groups which are examined on a before and after basis. Control and experimental groups have
been designated for each of the neighborhood communication centers which
are located in different kinds of residential sites. The evaluative
research allows service delivery to be analyzed in three contexts: two-
way cable; one-way viewing with telephone call-in; and those with no
access to the system at all.

In Spring 1975 implementation and research efforts were initiated in Reading. This involved hiring and training local staff, selecting and installing equipment, and identifying and contacting various community organizations. Simultaneously, the design of evaluative instruments was undertaken by both the research team and social service agencies. Programming began in January 1976. By that time, many senior citizens, agency-users, and representatives of local organizations had acquired familiarity, skill, and experience with the technology.

The design and implementation of this experimental cable project have been the product of three criteria: the needs of senior citizens, the technical configuration of the cable system, and the requirements of evaluative research. The strong and active involvement of senior citizens and local community groups has been an essential component of this project. Senior citizens participate in virtually all aspects of the interactive cable system from planning to actual production. They are responsible for planning and developing software and are involved in the operation of the neighborhood communication centers and technical equipment.

A diversity of public and quasi-public organizations use the interactive system to communicate with senior citizens. More than seventy agencies have participated in the programming. Twenty organizations are regular users of the two-way cable system and fifty have appeared on an occasional basis. Educational institutions account for 15 percent of the organizational programs, local governmental units comprise 21 percent of such programming, and social service delivery agencies are responsible for 49 percent of the programs.

At the present time, the two-way cable system is operated by a local non-profit corporation, Berks Community Television. This corporation was created to take over the interactive cable system at the termination of the experimental phase in February 1977. It represents a broad range of public and private institutions in the local community.

The interactive cable system has generated a wide array of effects on both individuals and organizations in Reading. Senior citizens constitute 16 percent of Reading's total population and are significant consumers of public services. Moreover, they often face substantial problems of limited mobility and access in obtaining the services they require. The two-way cable system has emerged as a means for the elderly to communicate with each other and with public agencies. Furthermore, it serves important social and political purposes by reducing isolation and providing a forum for the elderly to participate in local government processes.

One area in which programming over this system has proven to be particularly effective is that of citizen-government interaction. Regular weekly programs are conducted in which senior citizens can communicate directly with elected municipal and county officials. Senior citizens utilize these programs to articulate their preferences about public goods and services provided by local governmental units. Requests for information, specific demands, and evaluations of municipal policy are made for such local issues as street repair, water supply, housing, property taxes, and safety.

These programs allow senior citizens to participate directly in public affairs without encountering the time and travel costs of visiting City Hall or the institutional and psychological constraints of participating in formal public meetings. The interactive cable system personalizes the contact between citizen and public official and enhances the traditional functions of local government officials. Elected officials are able to obtain accurate and regular information on citizen concerns without leaving their offices, and can also use the cable system to explain the constraints and dilemmas they face in resolving urban problems. Thus, the overall level of information about urban conditions is increased for both citizens and public officials.

The character of citizen-government interaction on the Reading cable system is in sharp contrast to the type of interactive telecommunications most frequently proposed for use in local political processes: polling and referenda through digital return systems. The experience in Reading indicates that regular teleconferencing between citizens and public officials should be seriously considered as a mode of communication in local political processes. It allows citizens to transmit their demands for public goods and services to local decision-makers on a regular basis and permits policymakers to obtain first-hand information from consumers of local services. Moreover, such teleconferencing can be readily incorporated into the structure of urban government, for it compliments existing communications processes and does not entail extensive modifications in the daily operations of local government.

Social service programming consists of information exchanges between senior citizens and representatives of the Social Security Administration, County Board of Public Assistance, and other social service agencies. In addition, information on health care and medical services is transmitted on a regular basis. Adult education programs are conducted by the local branch of Pennsylvania State University and several high schools use the cable system for social studies classes in which students discuss civic affairs with senior citizens.

A wide range of entertainment programs take place over the two-way cable system. Senior citizens have created quiz shows in which homeviewers and participants at the neighborhood communication centers compete for prizes. In addition, there are poetry readings, sing-alongs, and programs which highlight the local history and folklore.

There are approximately 35,000 subscribers to the local cable company, ATC-Berks TV Cable Company. A survey of subscribers was conducted in

November 1976, just one month after programming had been made available
to the entire community. Nearly one-fourth of all subscribers in the
survey reported that someone in their household had watched the interactive cable programming. Moreover, almost half of the cable subscribers
were elderly or had an elderly person living in their household. These
findings indicate that there is a large potential audience for such cable
programming in urban communities oriented toward the elderly.

The interactive cable system in Reading has provided a human and technological framework for the development of a broad range of public service programming. Public officials, service agencies, and senior citizens are able to determine if and how two-way cable television can serve
their particular needs and then design programs which are conducive to
their communication requirements and processes.

The Reading system does not depend on elaborate technological equipment
but rather on the capacity of individuals and organizations to formulate
new uses of two-way cable television within a community based and operated system. The Reading system encourages the development of a variety
of new applications. The 'open-ended' character of the system has led
to the rapid diffusion of this technology to other governmental units in
the community.

The findings of the Reading cable project indicate that public agencies
utilize interactive telecommunications for a variety of reasons; however,
production efficiency in the provision of goods and services is rarely
the basis for organizational participation in the cable system. Public
sector organizations obtain a variety of benefits from participation in
the two-way cable system. Certain agencies regard the interactive cable
system as an innovative means of providing outreach services which are
otherwise conducted through staff visits to individuals and community
centers within the urban area. The municipal and county governments
view the cable as a mechanism for obtaining citizen feedback on public

policies and programs, while other service agencies utilize the two-way programming to disseminate information to clientele who are tradition-ally hard to reach. And, for some organizations, the system serves as a valuable tool for gaining exposure and enhancing organizational status in the community.

In the past, a number of factors have limited the application of new technology in local government. A critical factor in the success of the Reading project is that the two-way cable programming does not conflict with the traditional norms and practices of the service delivery organi-zations. It is clear that, if two-way cable television is to be adopted by local units of government, it must correspond to the current needs and values of service delivery organizations.

For the elderly, the two-way cable programming has only partially led to greater utilization of social services. A far more pervasive set of effects has been in the social psychological health of the senior citizens and on their relationships with local officials and the com-munity at large. The two-way cable programming serves as an important vehicle for social interaction and participation in community affairs. It has provided a means for improving the responsiveness of public of-ficials to the distinctive needs and interests of the elderly population. Moreover, the interactive cable system personalizes the communication between citizens and local officials. Telecommunications eliminates the impersonality of bureaucratic structures and thus humanizes the relation-ship between the individual and the government.

The senior citizens in Reading have also developed a high degree of personal efficacy through their involvement in the cable system. Their access to, and control over, the public service programming has enhanced their overall visibility in the community and provided a forum for com-municating with each other in a collective setting. Furthermore, the participation of senior citizens in the planning, production, and programming has created a situation in which their skills and experience can be effectively utilized.

The results of this study illustrate the capacity of interactive tele-communications to reach specialized groups within large central cities. The Reading project provides a multiplicity of services over two-way cable television to one distinct population group: senior citizens. The elderly represent a growing segment of the population which can be effectively served by broadband communications. They have distinctive problems of isolation, limited mobility, and particular needs for in-formation and social services.

The potential abundance of channel space and the capability of two-way communications suggest that the needs of other subgroups of the urban population may also be addressed by new applications of cable television. The challenge for telecommunications is to create an institutional frame-work in which individuals and groups can make efficient and effective use of two-way cable television. Such a framework should allow a wide range of individuals and groups to participate in the planning and de-velopment of public service programming.

New public uses of two-way cable television are most likely to succeed where they reflect the needs and interests of local citizens and agen-cies. A participatory framework for testing and implementation is nec-essary for the development of innovative and imaginative uses of two-way cable television. Such a framework creates a self-generating process in which the full potential of interactive telecommunications can be realized in urban communities.

Dialogsysteme für den Einsatz im lokalen Bereich

Diese Arbeit berichtet über ein von der New York University in Reading, Pennsylvania, durchgeführtes Experiment, das dazu diente, die Auswirkungen des Einsatzes von Zweiweg-Kabelfernsehen bei der Bereitstellung von Diensten für ältere Menschen zu beurteilen. Dabei werden die Auswirkungen, die das Dialog-Kabelfernsehen auf ältere Menschen und Bereiche des öffentlichen Dienstes ausübt, beschrieben und analysiert. Die Arbeit behandelt die Faktoren, welche die Anwendung des Kabelfernsehens durch Organisationen im öffentlichen Dienst beeinflussen und untersucht die Notwendigkeit, entsprechende institutionelle Vorkehrungen bei der Entwicklung von Dialog-Kommunikationssystemen für den Einsatz im öffentlichen Bereich zu treffen.

Overview of CATV Developments in the U.S.

William F. Mason
McLean, Virginia, USA

ABSTRACT

During the 1960's, a great many creditable predictions were made about the use of two-way cable television systems to provide a wide variety of public and commercial services into private homes and between all kinds of public and commercial institutions. These capabilities have not been realized, but a number of interesting programs have been implemented to demonstrate the potential. This paper identifies and summarizes the most significant studies of the potential of cable, suggests the reasons that two-way cable systems have not been commercially attractive, and reviews the status of two-way systems in the United States.

Background

The development of cable television systems technology and cable equipment capabilities in the United States has far outpaced the development of the services and programs that might make use of such systems.

In the mid-1960's, it was realized that the many cable systems that had been installed in the United States were capable of providing a great many more services than they were actually delivering. The engineering community was very much aware of the fact that the cable systems being installed could provide 300 Mhz of bandwidth into homes, even though less than 75 Mhz was then being used. Although there was no public demand for increased services, speculation about the tremendous potential of cable began to be popular.*

Table I, "Milestone Events in Recent CATV History," lists some of the key events that mark the rapid growth of interest in cable between then and now. The first significant event was in 1969 when equipment suppliers assembled a report pointing out the potential of cable and recommending that the Federal Government begin to plan for wideband capabilities in all homes. A popular article called "The Wired Nation" appeared

---

*
1. Friendly, Fred W., Chairman. Report to Mayor John V. Lindsay, New York, N.Y., Mayor's Advisory Task Force on CATV and Telecommunications, 1968.
2. Rostow, Eugene V., Chairman. Final Report of the President's Task Force on Communications Policy. Washington, D.C., U.S. Government Printing Office, 1968.

TABLE I

MILESTONE EVENTS IN RECENT CATV HISTORY

1969   Electronics Industry Association, The Future of Broadband Communications, recommended FCC planning for cable.

1970   Ralph Lee Smith, "The Wired Nation," Volume 210, No. 19, The Nation, May 18, 1970. This report popularized the subject.

1970   Robert W. Peters, Charles E. Erickson, and Dieter Lohr, Business Opportunities in Cable Television, Volume II: The Outlook for CATV Equipment and System Construction, SRI Project 1C-7978, Stanford Research Institute, Menlo Park, Califronia, September 1970.

1971   Sloan Commission Report, On the Cable, The Television of Abundance, ISBN 0-08-058205-X (pbk), McGraw-Hill Book Company, 1971. This report gave intellectual support to the possibilities.

1971   The White House Office of Telecommunications Policy, Pilot Projects for the Broadband Communications Distribution System, Contract No. OTP-SE-72-105, Malarkey, Taylor & Associates, Washington, D.C., November 1971. This report presented plans for a large-scale demonstration of all proposed services.

1972   William F. Mason, et al., Urban Cable Systems, M72-57, The MITRE Corporation, Washington Operations, McLean, Virginia, May 1972. This report was the first complete design of a complete wired-city system and was prepared under a grant from The John and Mary R. Markle Foundation.

1972   Federal Communications Commission, CATV Report and Order, Operational Rules and Standards, February 1972. This Special Report required local government and community services channels and two-way capabilities be provided on new systems.

1975   Federal Communications Commission, Cable Television Technical Advisory Committee Report to the Federal Communications Commission, FCC CTB 75-01, May 1975. This report details the technical standards required for a wideband two-way CATV system.

1976   U. S. Department of Commerce/Office of Telecommunications Lowering Barriers to Telecommunications Growth, OT Special Publication 76-9, November 1976. This report speculates on how to make wideband capabilities replace other means of providing the many services that have been suggested.

in 1970 describing cable systems and their potential in terms that the public could appreciate. It made glowing statements about the profitability of cable systems and the money that could be made with two-way services. It also speculated on the power that could be exercised over a national network of such systems and proposed that the Federal Government, specifically the Federal Communications Commission, establish controls and procedures to ensure that the new communications media would be used in the public interest.

The country's larger industrial organizations took note and, later in 1970, the Stanford Research Institute contracted with a group of these industrial corporations to define the commercial possibilities of such broadband systems. The SRI study, "Business Opportunities in Cable TV," was proprietary to the clients who sponsored it, but its projections concerning the profitability of computerized two-way data systems and other markets for broadband communications had a significant affect on the government's interest in cable TV.

Following a 1970 decision by the Alfred P. Sloan Foundation to establish a Commission on Cable Communications, a number of studies were funded and assembled in a 1971 report called, "On the Cable, the Television of Abundance." This book, which sold widely, included some highly respectable projections on the prospects and possibilities for entertainment, the influence of cable on regular over-the-air television, the influence of cable on other news and public services media, politics, the community voice and the need for regulation. It did much to promote federal interest in the idea of government influence on the growth of cable TV - so that it would evolve with a "public interest" circulation.

In 1972, the Markle Foundation funded The MITRE Corporation to do a cable system design for large cities as a first step towards developing a real "wired city." In the same year the FCC published its famous "CATV Report and Order," and its "Operational Rules and Standards for CATV." This latter required that new cable systems provide more channels, some of which had to be dedicated to community services and educational capabilities.

In parallel with these landmark efforts, the cable industry began to advertise its ability to provide more than 20 channels of broadband communications into every home and capabilities to provide burglar alarm and two-way services. The public responded and cities began to require that new services be provided as a condition of licensing (franchising) cable operations in their cities.

Many federal agencies established a point of contact for cable matters and a number of "demonstrations" were developed, some under federal funding and some by large cable companies, to show the potential of cable. These demonstrations, some of which are mentioned in the following sections, showed how the home television set could be used to get all kinds of public information and to perform transactions. The National Science Foundation has summarized a number of these CATV demos under such headings as Citizen-Government, Interaction, Community Dialogue, Community Information Center, Counseling, Education, Employment, In-Service Training, Library Service, Management Facilitation, Public Information, Public Safety and Specialized Services.*

In 1971 the White House Office of Telecommunications Policy undertook to define "a major telecommunications pilot program to determine the usefulness and economic viability of wideband distribution facilities in alleviating some urgent problems of today's society." In coordination with such major U. S. Agencies as HUD, HEW and the Department of Commerce, ideas for a large-scale demonstration were considered and in November 1971, OTP issued a report "Pilot Projects for the Broadband Communications Distribution System," which examined the various experiments, demonstrations and candidate cable systems that might provide the capabilities needed for a new large-scale demonstration and presented a demonstration plan. The planned services included merchandising, electronic banking and credit authorization, traffic monitoring and control, municipal fire and police call box services, utility meter reading and automatic load control, home, school and small business security and automatic vehicle location, education, medical and health services, commercial, polling and channel monitoring, data and library retrieval capabilities, income tax return filing by cable, and experiments with cashless transactions, employment services and home video jukeboxes.

No Market for Two-Way Services

Why have the exciting prospects for cable not been realized? The reasons are fairly straightforward, having to do with the nature of the cable TV business and the way the U. S. Government reacts to new ideas.

The cable industry has derived from the efforts of a very large number of small business entrepreneurs who built very profitable cable systems in areas where over-the-air reception is bad. Using community antennas requires only a modest investment in facilities to meet an extremely strong demand. The situation was perfect for making money. Recognizing this, a number of large companies, called "multiple system operators," bought up many of the small cable systems. The accent was still of course on profit and, while the quality of service improved somewhat, the number of services was only improved in cases where they could attract more subscribership. Any "service" that required more than a video tape, in a broadcast mode, was considered "blue sky" -- the code word for ideas that are not commercially attractive.

---

*National Science Foundation, Social Services and Cable TV, NSF Grant No. APR75-18714, July 1976.

TABLE II

FAMOUS TWO-WAY EXPERIMENTS*

AMHERST, NEW YORK	Interactive to Handicapped
DENNIS PORT, MASSACHUSETTS	Switchable Video Origination Points
JONATHAN, MINNESOTA	Community Information Center
LOS GATOS, CALIFORNIA	Early Prototype Equipment Tests
ORLANDO, FLORIDA	Variety of Retrieval Services
OVERLAND PARK, KANSAS	School Services to Homebound
RESTON, VIRGINIA	Many Advanced Service Concepts

NEW TOWNS WITH BROADBAND COMMUNICATIONS:

ROSSMOOR'S LEISURE WORLD, MESA, ARIZONA

ROSSMOOR'S LEISURE WORLD, COCONUT CREEK, FLORIDA

FLOWER MOUND NEW TOWN, DALLAS, TEXAS

WOODLANDS, CONROE, TEXAS

GATEWAY, ALBANY, CALIFORNIA

TAMA, JAPAN

HIGASHI - IKOMA, JAPAN

---

*There are other systems which included plans for two-way capabilities, but which have not been operational with noteworthy demonstrations. These include: Akron, Ohio; Welfare Island, NYC; New York City; St. Charles, Maryland; Pensacola, Florida; Disney World, Florida; El Segundo, California; Bethlehem, Pennsylvania; Peoria, Illinois; Irving, Texas.

Indeed the many studies that analyzed what it would cost to deliver interactive or public services on TV all concluded: the additional services would not be profitable unless they could be aggregated into a relatively large group. Even then, public monies would have to be used, rather than straight subscriber fees, because the public is not interested in paying the extra costs for the new things that have been demonstrated so far -- except for games. A new market for computer games like PONG was spawned, using microprocessors in TV sets rather than a central computer.

On top of the financial dilemma is the problem of 100% subscribership. Many of the services that interest municipalities would require that all homes be served by cable. One hundred percent "penetration" would be needed. Only then could the conventional means of delivering the particular services be eliminated. For example, electric or gas meter reading by cable would not be practical unless the present manual system could be abandoned. Since most cable systems produce maximum profits at much less than 100% penetration (due partially to the high cost of serving some hard to reach homes) -- there has been no motivation to achieve 100% subscribership.

Another problem has been that some attractive "blue sky" data retrieval services would require that cable systems be interconnected, but these particular services would not justify the interconnector costs. Although many analysts feel we have a "chicken and egg" situation, i.e., if the services were operational, enough demand would be generated to make them profitable, the entrepreneurs remained skeptical -- and rightfully so.

Although some additional one-way services were added when the FCC required new systems to provide additional channels, interactive TV has not been made available. The closest we have to such services are the several computer-aided instruction systems operating in colleges and a number of electronic funds transfer or banking arrangements that can be exercised through the telephone system.* Wideband and video are used in a limited number of computer to computer arrangements within specific companies and the rural health programs of HEW continue to fund experiments providing physician support to remote clinics, but in a very real sense, no cable system is exercising general purpose wideband or interactive services as of this date (1977).

Table III illustrates these points. From a MITRE analysis of three types of systems for Washington, D.C. (see Urban Cable Systems reference in Table I) one can see that the return on investment for a sophisticated two-way system is not anticipated to be higher than for a more simple system, and the capital risks are much higher. Figure 1, from this same analysis shows how profitability drops when one tries to get 100% penetration. The lower subscriber rates required, plus the high cost of serving the most difficult-to-reach homes makes 100% penetration impractical from a businessman's point of view.

---

*National Science Foundation, The Consequences of Electronic Funds Transfer, NSF C844, June 1975.

TABLE III
SUMMARY COMPARISON OF THREE SYSTEM OPTIONS
Source: Urban Cable Systems

SYSTEM CHARACTERISTICS	ONE-WAY SYSTEM	TWO-WAY, WITH SUBSCRIBER RESPONSE SYSTEM	TWO-WAY ELECTRONIC INFORMATION HANDLING SYSTEM
Number of Channels Telecasting			
Total	up to 30	up to 64	up to 64
Forward	up to 30	up to 60	up to 60
Reverse	4 (SA, 2, 7 only)	4	4
Public Access	12	24	24
Imported TV and Cable Signals	5	6	6
Point-to-Point Total	150	150	150
Number of Two-Way Terminals			
Total (SR/Combined)			
1st Year	100/0	100/0	100/100
5th Year	6,100/0	121,936/0	121,936/73,165
10th Year	6,100/0	204,124/0	204,124/112,474
Households Passed, 10th Year	262,796	262,796	262,796
Final Penetration	76%	78%	78%/47%
Capital Expenditures, by 10th Year	$31.2M	$61.2M	$114.0M
Operating Expenses, 10th Year	$6.72M	$11.1M	$20.1M
Per Subscriber Per Year	$34	$54	$99
Programming Costs 10 Years	$15.0M	$34.3M	$58.5M
Installation Fees (65% of Sub.)	$10.00	$10.00	$10.00
Operating Fees			
1st Set, Final Fee	$3.50/Mo./Sub.	$6.50/Mo./Sub.	$22.00/Mo./Sub.
2nd Set	$2.00/Mo./Sub.	$2.00/Mo./Sub.	$ 2.00/Mo./Sub.
FM Radio	no charge	no charge	no charge
Rates for Leased Channels	$20/Hr/Channel	$20/Hr/Channel	$20/Hr/Channel
Rates for Advertising	$200/Min/All Chan.	$200/Min/All Chan.	$200/Min/All Chan.
Rates for Utility/Maint. Services	–	$2.00/Mo./Hld.	$2.00/Mo./Hld.
Rates for Leased Point-to-Point Channel	$4/Hr/Channel	$4/Hr/Channel	$4/Hr/Channel
Gross Revenues, 10th Year	$11.5M	$26.3M	$49.2M
Per Subscriber Per Year	$58	$128	$240
Equity Required (Debt/Equity = 2)	$9.9M	$18.7M	39.2M
Loans Required by 10th Year	$22.8M	$35.6M	77.9M
Interest Rate on Loans	8%	8%	8%
Break-Even Points			
Operating Income	4th Year	4th Year	4th Year
Pre-Tax Profit	7th Year	4th Year	4th Year
Coverage of Interest			
Payments	6th Year	4th Year	4th Year
Minimum Ratio of System Assets-to-Outstanding Debts (min. Year)	1.19 (Year 6)	1.13 (Year 3)	1.07 (Year 3)
Estimated Value of System, by 10th Year	$22.7M	$115.8M	$242.6M
Rate-of-Return on Equity 10th Year	11.9%	30.2%	28.4%
Rate-of-Return on Investment 10th Year	0.0%	10.8%	10.3%

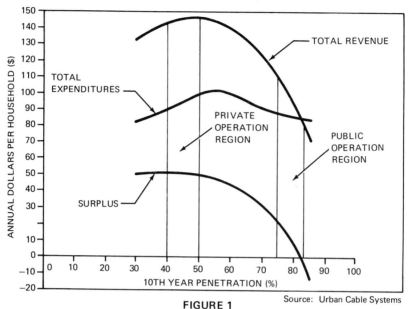

FIGURE 1
REVENUES, EXPENDITURES AND SURPLUS VS. PENETRATION
FOR SYSTEM WITH TWO-WAY SERVICES

The Current Situation

In January 1976, there were approximately 3,400 local cable systems serving around 10,800,000 U.S. households. Almost a billion dollars of annual income was being generated each year in subscriber fees on a plant investment of about $1B, but the rapid growth of the industry had greatly slowed down from about thirteen percent per year in the early 70's to two percent in 1976. Although many two-way systems have been "demonstrated", only a very few remain operational. The remainder of this paper will categorize and summarize these surviving two-way cable systems.

Urban Administration

Urban Administration includes police, fire, traffic and city administration.

- Columbus, Ohio has a traffic control and surveillance system.

- Philadelphia, Pennsylvania and Phoenix, Arizona use video teleconferencing in their law enforcement and criminal justice systems.

- The New York City Metropolitan Regional Council operates an urban administration conferencing system involving 13 different locations.

- Spartensburg, South Carolina trains day-care workers and Rockford, Illinois trains firemen on two-way cable.

Hospitals and Health

Veterans Administration Hospitals now include specialized broadband services. Telemedicine experiments have been set up in seven locations in the United States by the Department of Health, Education and Welfare. These locations are listed in Table IV.

Industrial

General Motors, American Motors, Dow Chemical and Rockwell International all use two-way broadband systems in their factories to improve inventory and quality control surveillance, security alarms and training.

Commercial

A number of banks and very large commercial buildings use cable to exchange computerized information within their own complexes.

Military Bases

The Army is testing cable systems for administration, training, security and containment on military bases. The Navy has TV distribution capabilities on 160 ships, 138 of which have additional broadband shipboard information training and entertainment capabilities.

Education

Computer-Aided Instruction systems are operating in a number of universities and other schools in the U. S. Most notable are the University of Illinois, Brigham Young University, Phoenix College and Northern Virginia Community College. In addition, there are dozens of other schools making use of video tape libraries that can be remotely ordered.

Social Services

The National Science Foundation has a number of teleconferencing experiments in institutions for the elderly and among day-care centers.

Summary

While the forecasts of the late 60's have not yet been realized, wideband capabilities are beginning to appear in institutional settings. Meanwhile the development of low cost microprocessors is making computer-based games available on home TV sets. Next year's microprocessors will offer even more services in the home and it now seems likely that any service that requires a two-way cable system will be even harder to finance than it might have been before microprocessors developed. We may yet see the "smart terminal" approach, networked in a cable system, but it now seems more likely that the telephone system will provide the universal networking once associated with "blue sky" predictions.

TABLE IV

CABLE TV TELEMEDICINE PROJECTS SPONSORED BY HEW*

Title	Health Specialty	Locations
Bethany/Garfield Community Health Care Network	Medical (hospital and clinic), laboratory, pharmacy, administration	Illinois • Bethany Brethren Hospital • Garfield Park Hospital • Bethany Clinic • May-Rosen Clinic • Bethany Drug Center
A Comparison of Television and Telephone for Remote Medical Consultation	Primary health care	Massachusetts • Cambridge Hospital • Fitzgerald School Adult Health Center • Donnelly Field Neighborhood Health Center • Neighborhood Family Care Center
Case Western Reserve School of Medicine Anesthesiology Project	Anesthesiology	Ohio • Case Western Reserve University • Veterans Administration Hospital
Illinois Department of Mental Health Medical Center Complex Community Mental Health Program Picturephone Network	Mental Health	Illinois • Illinois State Psychiatric Institute • Healy School • Pilsen Mental Health Center • West Side Organization • Mental Health Center
Lakeview Clinic Bi-Directional Cable Television System	Rural group practice	Minnesota • Lakeview Clinic (Waconia/Johnson) • Ridgeview Hospital
Mount Sinai Wagner Bi-Directional Cable Link	Pediatrics, Orthopedics, Psychiatry	New York • Mt. Sinai Hospital • Wagner Child Health Station
The Provision of Speech Therapy and Dermatology Consultations via Closed Circuit Television	Speech therapy, Dermatology	Vermont/New Hampshire • Claremont General Hospital • University of Vermont Medical Center • Dartmouth-Hitchcock Medical Center

*The MITRE Corporation, Benefits and Problems of Seven Exploratory Telemedicine Projects, MTR-6787, February 1975.

Überblick über die Entwicklung des Kabelfernsehens in den Vereinigten Staaten

In den 6Oer Jahren machte man eine ganze Reihe von bemerkenswerten Voraus-
sagen über den Einsatz von Zweiweg-Kabelfernsehsystemen zur Übertragung
verschiedener öffentlicher und kommerzieller Programme in private Haushalte
oder zwischen den einzelnen öffentlichen oder kommerziellen Einrichtungen.
Zwar wurden einige interessante Programme zu Demonstrationszwecken ver-
wirklicht, die erwähnten Betriebsmöglichkeiten jedoch nicht ausgeschöpft. In
dem vorliegenden Beitrag werden die wichtigsten Studien über entsprechende
Kabelanlagen genannt und kurz besprochen. Außerdem werden die Gründe für
die wirtschaftliche Unattraktivität von Zweiweg-Kabelsystemen untersucht
und der Stand solcher Kabelsysteme in den Vereinigten Staaten beschrieben.

The Design, Implementation and Operation of Bidirectional Wideband CATV Networks for Two-Way Video Audio and Data Services

Edward J. Callahan Jr.
Denver, Colorado, USA

ABSTRACT

This paper describes the design, implementation and maintenance of the broadband two-way CATV systems. These systems are used for distribution of video and audio signals, both one-way and two-way, as well as two-way data and interactive services. Several experiments will be described utilizing such two-way services. The unique system concept employing a hub design with radial trunk lines will be shown to be a most powerful concept in implementing these services.

American Television and Communications Corporation (ATC), is a CATV Multiple System Operator (MSO) headquartered in Denver, Colorado, and currently operates 97 systems in 31 states, serving nearly 600,000 subscribers. ATC was founded in 1968 for the purpose of providing broadband communication services. These services have taken the form of broadband two-way CATV systems, common carrier microwave systems, and most recently, satellite communication systems. ATC adopted the philosophy that in order to offer a viable package of communication services in metropolitan areas, we must have the flexibility that a bidirectional cable system offers. Every system we have designed, therefore, has had inherent in it two-way capability. Since 1970, ATC has designed and built over 3500 miles (5600 km) of two-way CATV plant.

When we speak of two-way cable TV, we are speaking of a total system concept. This concept is not simply a matter of placing amplifiers every 22 dB of cable, but must be a well planned design with close attention to the total system for consistent optimum service. In CATV the service is multi-channel distribution of video and audio services signals. The "nucleus" of our system is the "head-end" where off-the-air broadcast signals as well as microwave signals are received and processed and delivered to the cable system. In addition, premium or pay television signals either locally originated or received from a domestic satellite are integrated here. Our system, therefore, implies more than simply installing cable and amplifiers and hoping that the desired result is obtained.

Our "system" is the total path from the signal source in the previously
described head-end all the way to the furthest subscriber at the very
end of the distribution network. This system must be transparent. This
means that no visible degradation is added to the signal from the time
it leaves the head-end until the time it reaches the most distant subscriber.

The hub concept which we employ in our broadband bidirectional cable
systems has proven to be a most powerful concept. The trunk lines ema-
nate out from the hub in a radial fashion like the spokes of a wheel.
The reason for the radial distribution is that short amplifier cascades
are permitted, typically 20 or fewer amplifiers. This is important because the reliability of service is enhanced since the total population
of subscribers on any one trunk line is limited. Therefore, in the event
of an equipment failure, fewer subscribers are affected. Also, electronics are located where they are accessible so that the mean time to
repair, in the event of a failure, is minimized. If more than one hub
is required because of the geographical size of the community to be
served, a general rule of thumb is that a cable two-way interconnect
will be used if there are only two hubs; and microwave interconnect will
be used if there are more than two hubs at present, or if there is the
future possibility of more than two such hubs being interconnected.

Now that the hub concept has been established as the overall design for
the plant, let us take a look at system distortion parameters. We appor-
tion system distortion between the trunk system and the distribution
system. We allow the trunk system to be limited by signal-to-noise
establishing a 47 dB signal-to-noise ratio at the input of a multi-
channel converter. The distribution system is limited by signal dis-
tortions with the allowable cross-mod ratio being 52 dB. The second
order harmonic ratio is 60 dB and the intermodulation distortion ratio
is 52 dB. Once the system distortion parameters are determined, the
next step is to determine the number of channels required now and in the
future. If a system is initially activated with almost all channel capacity utilized, then one must ask if a second cable should be installed
along with first cable, thus, giving twice the potential capacity of a
single cable if installed on both trunk and feeder. Another alternative
is to install a single cable initially and to go back at a later time
and overlash or install a second cable when the total capacity of the
first cable has been occupied. The time value of money is obviously a
consideration here.

To insure the transparent signal transportation system previously mentioned, we specify the use of hybrid integrated circuit (IC) amplifiers with the use of automatic gain and level control at every station. This enables us to essentially maintain the controlled AGC window down the feeder to the furthest subscriber. I am stressing the concern of our company that strict level controls are maintained throughout the entire system. We feel that the transparent system concept we employ gives us the quality control over the product we deliver to maintain the high standards necessary to insure that our subscribers get the best service possible -- efficiently, economically, and with a minimum of interruptions.

Now let's look at the mechanics of the design that enable us to accomplish the above-mentioned goals. The first step in having an accurate design is to insure that the strand maps or maps of the city streets over which the cable will be installed are completely accurate. Therefore, every foot (meter) of every street in which cable will possibly be installed, is wheeled off with devices that are accurate to one foot (.305 meters). These maps show all available strand connections and all potential subscriber locations. Once this information is assembled, we then use a ATC designed program for the Wang 2200 Computer system. This system employs a CRT display, a cassette tape reader and a keyboard in Basic II language to insure a workable two-way system design. The reverse path calculations are made simultaneously with the forward path calculations so that the proper signal levels will be attained back at the hub location.

Our design follows the logical flow of the signals. The lowest cost area of design occurs when no trunk cable or amplifiers are used, therefore, only feeder cable and line extenders are used in dead-end pockets. Amplifiers are placed in locations where the designer can go in as many directions as possible with the output signal. This is accomplished by the use of passive splitting devices. The tedious calculation and selection of these devices is aided by the computer. The amplifier location is backed out of the area until a signal has been used to its maximum possible level. Along the cable are directional tap off devices at each utility pole for every potential subscriber in town. Again, the computer chooses the values of these devices. Once the line extenders have been positioned in optimum locations, the bridger amplifiers are properly positioned to serve the line extenders. A bridger should be on or near a main street since it will be served by the trunk line. The final step in design layout is to connect all the bridgers with trunk cable and trunk amplifiers.

The first actual implementation of these design practices was the system
constructed in Orlando, Florida. The intent of the design in Orlando
was a multi-hub configuration serving up to 18 distinct communities.
The system in each community was to have bidirectional capability so
each community would have return capability to the communications central
in Orlando. When the initial portion of the Orlando system was activated,
it was discovered that the reverse path was being degraded by multiple
points of RFI ingress. Through a series of carefully controlled engineering experiments, it was discovered that the television set, the subscriber drop and the connectors in the trunk and distribution system all
allowed ingress into the system. The solutions for these problems included a high pass filter transformer at the subscriber's television
set, a well-shielded drop cable and RFI sleeve connectors in the trunk
and distribution system. The RFI sleeves in the connectors eliminated
the necessity for going back on a regular basis and retightening connectors to prevent either radiation from the system or RFI ingress into
the system.

The maintenance of such a two-way system required several new techniques
to be developed in regard to the problem of ingress to insure that the
system met its performance specifications over a long period of time.
Several methods of determining ingress into the system were tried. One
of these was to use a very low power portable transmitter while walking
along the cable system until a point of ingress was determined. The
detector was usually a spectrum-analyzer located at the nearest hub to
the point of ingress. Another method was to modulate an FM carrier with
unique modulation and use a portable FM receiver to locate radiation
from the system. This was found to be very effective.

Uses for Bidirectional Cable Systems.

Several uses for the reverse path have developed in the Orlando, Florida
system. One of these is the use of coded signals transmitted from the
office location back to the head-end to control the non-duplication
program switcher. Another use is to carry a high quality FM modulated
video signal which originates from a series of automated cassette tape
machines back up to the head-end where it is distributed out over the
Orlando system and also distributed through ATC's common carrier micro-
wave system to the east coast of Florida.

A full two-way interactive system was also installed in Orlando in 1972 for the purpose of demonstrating a two-way computer driven interactive system known as Polycom. The Polycom system philosophy was a polled interrogation and response system. The terminals were polled by the central computer and responded back to the central computer with their address and status word and data word if required. These terminals provided an interactive keyboard for such applications as opinion polling, catalog shopping, interactive educational courses and similar applications. The terminal also provided two channels of interdicted premium television programming.

The terminal had associated with it a junction box which permitted sensors to be connected to it for such applications as fire and burglar detection and also provided contact closures for applications requiring control signals, for example, the control of a video tape recorder or synchronization of a single frame video storage device. Additionally, there were digital interfaces provided to allow such things as facsimile printers or other data terminals to be connected to the interactive computer system. Twenty-five subscriber locations were equipped with these terminals and the technical feasibility of the hardware and software system design was clearly demonstrated. The system test continued for approximately one year and was terminated at that time because of the lack of economically viable subscriber hardware.

Another two-way system was operated in Reading, Pennsylvania to provide for the delivery of two-way audio-video services to the elderly. To maintain this program the New York University Reading consortium was formed. The consortium was comprised of the Alternate Media Center and the Graduate School of Public Administration both of New York University, the city of Reading and Reading Housing Authority, the Berks County Senior Citizen Council and Berks TV Cable Company, division of ATC. The experiment was funded by the National Science Foundation.

The implementation of the experiment was begun in June 1975 and entailed the establishment of three interconnected neighborhood communication centers. All the centers were equipped with the necessary hardware for receiving and originating signals and were interconnected via the two-way cable system enabling simultaneous two-way audio-video interaction between the centers. To augment the communication centers, converters were installed in 128 private senior citizen homes thus allowing these home viewers to receive the experimental programming and participate in the interaction via telephone. Using a midband Channel G, the experi-

mental programming began in January, 1976. The initial programs acquainted participating seniors with the operation of the interactive
system and they served as a means to explore areas of interest and concern to seniors which could be effectively dealt with via the interactive
system.

A year after its implementation the system had 26 separate programs all
produced by senior volunteers. Additional remote origination sites were
tied into the system, for example, the city hall, the county courthouse,
the social security office and the local schools. These additional locations facilitated participation in programs when time, distance or busy
schedules might otherwise preclude onsite participation. Examples of
some of the interactive programs include interactive discussions between
the three centers and the mayor of Reading, who is located at city hall;
interactive discussion between the communication centers and the Berks
County Commissioners who are located at the courthouse; verbal history
courses between the communication centers and students at local high
schools; discussion on current critical issues between the communication
centers and students at local high schools; discussion on social security benefits involving communication centers and the staff members at
the social security office. The program gives elderly persons and shut-
ins access to any number of service and educational programs which would
not be otherwise possible. In September, 1976, the programming was
moved from Channel G, a midband channel, to cable Channel 3 and became
known as Berks Community Television. This move was made in response to
a survey which indicated a 70% viewership among the participants. This
change broadened the participatory two-way nature of the program because
it could now be viewed by all cable subscribers who in turn could participate via telephone while the original group still had two-way audio -
visual interaction. The experiment which has helped to alleviate lone-
liness provided an avenue for reinvolvement, broadened the scope of
many lives and generally improved the quality of life for many participants, ended in February, 1977. Both the New York University involvement and the National Science Foundation financial support ended. The
interactive system has now reverted to the community. The Berks Cable
Television Advisory Board has been preparing for this changeover since
its inception in September, 1976 and they have been actively pursuing
community support and funding as well as developing ideas to improve
current programming and expanding the concept to include many other
facets of the community.

Data Applications Via Two-Way CATV.

Several applications for two-way data transmission for cable systems
have appeared in the past several years. Coaxial cable two-way systems
offer several advantages for data transmission. Since the cable system
is well-shielded from RFI, it provides a much less hostile environment
then does the traditional twisted pair communication circuits. It also
offers great flexibility in configuring various data channels since only
that bandwidth necessary for a particular service need be assigned to
it.

Here again the hub becomes very important in the overall effectiveness
of two-way communications on cable TV. A data path is established from
the remote terminal location through the reverse path to the central hub
location. At the hub, the signal is frequency translated to the spectrum in the forward direction and sent out to the appropriate receiving
terminal. An important point to keep in mind in such a hub system, is
that while the standard cable television service is common on all trunk
cables, the remaining bandwidth available may be used independently on
each of the outgoing trunk cables for various data or lease channel
applications. This again is due to the great flexibility of the hub
system concept.

In several field experiments utilizing data transmission on two-way cable
systems, the error rates in the receive data have been found to be extremely low, again pointing out the added advantage of communications
within a well-shielded environment. In a major metropolitan system, the
data channel may be routed back to one hub location, transmitted in the
forward direction through an interconnect system to a remote hub location and then onto a data terminal located in that part of the system.
It is conceivable that a totally automated computer controlled data system
could be implemented in which varying data rates would be available at
the transmitting terminal location simply upon demand. The terminal
would send a request to the computer at the central hub specifying the
data speed or bandwidth of the channel required. The computer would do
the necessary system reconfiguration, indicate to the transmitting terminal that the path to the desired receiving terminal had been established
and that communications could begin. A similar procedure could be established so that the distant terminal would have an appropriate path established back to the original terminal thus the path would be completed in
a two-way fashion.

In the not to distant future, there will be examples of wideband data communications circuits in which remotely located data terminals will communicate with each other via interlinking systems comprised of two-way microwave channels and two-way cable TV channels. Once the feasibility of such an interconnect system is demonstrated, it will be a simple matter to utilize satellite transmit/receive earth stations to link distant cities, into such a network. It will be services such as these that will truly broaden the scope of two-way interactive coaxial based systems.

Thus we see how the two-way cable TV system offers great flexibility for providing two-way audio, video, and interactive data communications. This truly is the promise of cable TV systems and with continuing advancements and improvements in technology, it is our belief that this promise will soon be a reality.

Planung, Realisierung und Betrieb von Zweiweg-Breitband-Kabelfernsehsystemen für die Übertragung von Bild, Ton und Daten

Der vorliegende Beitrag beschreibt Aufbau, Errichtung und Instandhaltung von Zweiweg-Kabelfernsehanlagen. Diese werden zur Verteilung von einfach und doppelt gerichteten Fernseh- bzw. Rundfunksignalen sowie für doppelt gerichteten Daten- und Dialogverkehr eingesetzt. Mehrere Experimente mit derartigen Dialogdiensten werden beschrieben. Hinsichtlich der Realisierung des Breitbandnetzes hat sich eine Struktur als besonders zweckmäßig herausgestellt, bei der Unterzentralen verwendet werden, von denen aus sternförmig verlegte Hauptverbindungsleitungen die einzelnen Gebiete des Netzes versorgen.

North American Experience With Reverse-Direction Transmission in Broad Band, Tree Structured Distribution Networks

I. Switzer
Mississauga, Ontario, Canada

The present population of broad band communications systems in the United States and Canada were developed as cable television systems, or perhaps more accurately, community antenna television systems. The first ones were built around 1952 to provide television reception in "shadow" areas or at distances too great from the transmitting station for conventional private receiving antennas to be effective. These systems evolved as "tree structures" with their "roots" at the head-end or main antenna, where signals were received to be fed through the "trunk" cables into the "branch" distribution cables into the "stems" or service drops to the "leaves" or subscriber receivers. Early systems were extended branch by branch as service was demanded and as systems grew in financial strength. The term "tree structured" is very appropriate.

These systems were developed by enterpreneurs who generally had little technical knowledge, particularly in the general telecommunications field, and who had no intent other than satisfying the growing demand for improved television reception. The early cables were co-axial types, those being the most available at reasonable cost, and being quite effective for VHF television signal transmission. Amplifiers were derived from existing television IF amplifier designs and also, in some cases, from broad band pulse amplifier designs. In time, the quality of the co-axial cables improved and improved amplifiers appeared. As solid state amplifying devices became available these were adapted for cable television use and introduced into regular system use. Today's cable systems are functionally and economically adequate (but not necessarily optimum) for their present function - distribution of broadcast television signals.

In the early seventies the cable television industry in the U.S.A. had reached a plateau of economic development. Most of the "classic" community antenna situations had been developed. Nearly every community that had a reception problem also had a cable television system to serve it. The larger, more populous communities did not appear to need cable television because they had what the residents considered to be an adequate number of local television stations and no serious television reception problems. Newly introduced legislation restricted the "importation" of more distant stations in order to protect local television stations in larger markets.

American cable television entrepreneurs cast a covetous eye on these large population centers and looked about for means to create a service that could be sold in communities that had a variety of local stations and no particular reception problems. Additional broadcast type services were conceived (but not necessarily born) and the concept of "two-way" cable was developed. At the same time the name of the industry underwent a subtle change from "CATV - Community Antenna Television" to "Cable Television" to indicate that the business now provided television services of a nature other than broadcast reception.

The "two-way" function was developed on a frequency diversity basis. The normal forward functions were to be carried on at frequencies above 50 MHz while reverse direction communications channels were to be provided on frequencies below 40 MHz, allowing a 10 MHz cross-over region for the required filters. Some more enterprising operators started building such systems or converting older systems by installing amplifiers with the required filters. Some problems were immediately apparant.

Early operations had trouble with the filters. Every amplifier in the signal stream had to be fitted with filters at input and output and these filters then became part of the overall frequency response of the system. Small ripples in the pass bands of the high pass filters quickly accumulated into unacceptable ripples in the overall system pass-band. Some problems were experienced with group delay distortion at the filter edges affecting channel 2 transmission. These problems were gradually overcome as manufacturers introduced more sophisticated filter design and manufacture techniques.

Considerable problems were experienced with signal ingress. The reverse signal path was often contaminated by signals leaking into the cable. Amateur radio transmitters, and powerfull HF radio broadcasting stations were the most common source of problems. It was found that many cable connector designs in common use were inadequate in that they loosened their grip on the aluminum cable sheats over a period of time. The aluminum tended to "cold flow" and recede from the cable connector grip. Improved connector designs have overcome this problem but there were additional problems with equipment enclosures, subscriber service taps, drop cables and connectors on drop cable. Achieving and maintaining physical system integrity was, and continues to be a major problem. The systems comprise hundreds of miles of cable with thousands of service drops, much of it risking exposure to EM fields that seek to penetrate the cable shielding and affect signal distribution.

Noise turned out to be a serious problem. As more and more terminals designed
to provide reverse direction communication were connected to the system their
noise contributions added in parallel and created a considerable noise level at
the head-end where all of these reverse direction signals concentrated. The main
forward signal system might be compared to a water distribution system drawing
water from the head-end reservoir and distributing it throughout a community.
The reverse direction system may be compared to the sewer system which picks up
a little bit of noise from each home with a terminal and gradually concentrates
the noise as it moves back toward the "sewer outfall" at the head-end.

These technical problems are gradually being overcome. The RF integrity of
the cable system is being improved considerably and the cable television industry
is developing the devices and techniques needed to cope with the problem. This
is being done not so much to facilitate reverse direction transmission as to
expand the capacity of the forward direction system. As systems augmented their
forward capacity by using "mid band" and "super band" spectrum (108-174 MHz and
above 216 MHz) they experienced interference to these signals from fields outside
the cable.

A technique has been developed and applied in a few systems to overcome
the terminal noise problem. Addressable switches at distribution bridging points
switch in only those distribution branches required at any particular time, thus
isolating other branches and their associated noise from the system. Most reverse
transmission communications systems operate in a "polling" mode and the controlling
computer can also direct the set up of the branch switches to establish the
required reverse path as necessary to allow polling of any particular area. This
method is very effective in controlling noise build up due to many terminals in
parallel. It does prevent simultaneous communications with other parts of the
system and this may be a serious or prohibitive limitation for some reverse
direction communications applications.

The development of practical reverse direction systems has also been hampered
by the failure of system operators to find practical, paying applications for
such capability. The only economic application which has surfaced so far is
PAY-TV. PAY-TV is presently operated in a subscription mode with subscribers
paying a monthly fee for the service. The only systems operating on a "pay per
program" basis are the very few systems which have developed practical reverse
direction communications for that purpose. The National Science Foundation in
the U.S.A. has funded some experiments in other forms of social service using
reverse direction transmission and this symposium is hearing reports from all

three of those experiments. My own personal conclusion from visits to two of these experiments is that social service applications of these types will be practical only as marginal increments on systems being paid for by PAY-TV businesses. I am convinced at the present time that PAY-TV is the only economically important application adaptable to the limited reverse communications capability of present tree structured cable systems.

I do not find this to be an unexpected or disappointing conclusion. These systems were never designed as optimum two-way communications systems. Their use in this mode was strictly an afterthought. Although our present telephone system was originally designed strictly as a voice communications system, it has been adapted with reasonable success for other purposes. Fortunately it had an architectural design in the form of a star configured, switched network which was applicable for other communications purposes. Even so we are presently witnessing drastic changes in our telephone systems as applications other than voice communications grow in economic importance. I suggest that the non-optimum structure and technology of our present cable system works against large scale, successfull implementation of two-way communications in a very general form. I suggest that the rapidly evolving telephone system with its more generalized transmission and switching capability will become far more important as a communications medium than will our present style of broad band, tree structured co-axial network.

Erfahrungen in Nordamerika mit der Realisierung und Nutzung von Rückkanälen in Breitband-Verteilnetzen mit Baumstruktur

Es wird dargestellt, wie sich in den USA und Kanada aus Gemeinschaftsantennen-anlagen die heutigen Breitbandkommunikationssysteme entwickelten. Diese Systeme haben ein Koaxialkabelnetz mit Baumstruktur. Für die Vorwärtskanäle wird der Frequenzbereich 50-300 MHz, für die Rückkanäle der Bereich 5-40 MHz benützt. Die Arbeit behandelt die Schwierigkeiten und Probleme mit der Abschirmung und den Filtern sowie dem Rauschen, das sich an der Kopfstelle der Rückkanäle zu einem beachtlichen Pegel addiert.

Die Entwicklung der Systeme mit Rückkanal ist neben den technischen Problemen vor allem durch das Fehlen wirtschaftlich sinnvoller Anwendung behindert worden. Pay-TV ist heute die einzige wirtschaftlich bedeutende Anwendung. Dies ist jedoch nicht so überraschend, wenn man bedenkt, daß dieses Systeme nie als optimale Zweiweg-Kommunikationssysteme konstruiert worden sind.

Two-Way CATV Systems

Sylvane Walters
Scarborough, Ontario, Canada

I. Introduction - (Abstract)

One of the topics generating the most interest today in the CATV field is that of
"Two Way". Various claims are presently being made as to the services that are
possible with the advent of full two way cable system. Such things as home security,
audience polling, merchandising, local origination for specific groups from different
locations and many other things are already being used and there are more to come.

It is the purpose of this paper to discuss some of the aspects of cable systems,
which apply mainly to the upstream portion of a bi-direction (two way) system. It
will cover the major factors to be considered in the layout and design of
bi-directional cable systems. The presentation is not directed towards attempting
to "retr fit" bi-directional facilities into existing unidirectional cable systems
although many of the considerations would still be applicable.

II. System Characteristics:

Before considering the technical aspects of a two way cable system, it is necessary
to determine some of the non technical objectives desired of the system, and before
these determinations should be attempted, certain general characteristics of such a
system should be comprehended.

The first is that a cable system, by its physical nature is most efficiently used as
a means of disseminating information from a single point, or route to a large number
of points or users, and inversely. It is also most efficiently used as a means of
acquiring information from a large number of remote sources, and transmitting it
back to a single point. Another feature is the very real limit existing in the
amount of bandwidth available in a given cable system. It costs a lot of money to
manufacture the bandwith and, even more to utilize it properly.

Different approaches exist in cable TV systems:

a) In areas where cable TV service to home subscribers with no return is desired;
in single cable, one way system may be utilized.

b) Where a limited video or data return path is desired the single cable system
may be equipped with two way transmission capability either sub. or mid split.

c) In many instances, it is desired to have a more broadband return service than that afforded by a single cable system. This may be from potential sources of video programming in a community to the head end, or possibly from a head end to a hub. In these portions of the system a dual trunk system may be used.

Both the second and third approaches are utilized in our systems of Scarboro Cable TV/FM and Credit Valley Cable TV/FM which form a part of the CUC group.

The terminology of the sub and mid split are used in the industry to define the design approach of the CATV systems. A sub split system would cover the range from 54 MHz to 300 MHz in the forward direction and from 5 MHz to 30 MHz in the reverse direction.

On the other hand a mid split system would cover 174 MHz to 300 MHz in the forward route and from 5 MHz to 108 MHz in the reverse one.

The CUC (Scarboro) Limited is using the sub split approach to carry 4 video channels and has proved to be very successful in spite of all the limitations.

The Credit Valley Cable TV/FM has the capability to use the other approach of the dual cable system. This is the concept of the "Wired City" approach. A certain section of the community has been wired for two way communication over the cable system. There are actually two cables, which are buried, one of which is for the forward transmission, the other for the reverse.

III. Design Aspects of the Scarborough System:

The various design aspects or factors for a two way system are as follows:

3.1. Band width and spectrum utilization:

At the outset of the planning of the Scarboro Cable TV system only one video channel (a community channel) was required to be transmitted in the return path from the studio to the head. This was not a problem at all.

The studio located about 6 miles from the head end did have a direct feed through 16 stations of trunk amplifiers. The video information for this community channel was modulated to Channel T8 in the sub low, sent through the return feed back to the head end with only 4 sub low return amplifiers. Once at the head end, was reprocessed from T8 to Channel 10 where it was coupled to the VHF system and distributed through the forward direction to the subscriber like the other channels.

In 1975 more demands for the extent and variety for services were initiated. Such services included:

a) CHILDREN'S CHANNEL, where children's programs with a non violent nature only would be carried.

b) CONSUMER INFORMATION CHANNEL where news and tips would help the family to save money, eat better and educate.

c) MULTI-CULTURAL CHANNEL includes several foreign language programs to accomodate the many ethnic groups in Toronto.

These three video channels were required to be added to the first. The allocation of four video channels in the sub low band was difficult because one of the first considerations in the development of a sub VHF return system is the allocation of frequencies within the available bandwidth. All CATV equipment manufacturers recommend the usage of only two video channels in the sub low, T8 and T9 and to leave the rest for data transmission. The only route we had to take was to carry the four video channels on the band and correct for any occuring problem.

CHANNEL	FREQUENCY	USE
T7	5.75 - 11.75 MHz	Video (Multi-Cultural)
T8	11.75 - 17.75 MHz	Video (Community)
T9	17.75 - 23.75 MHz	Video (Consumer)
T10	23.75 - 29.75 MHz	Video (Children)

This breakdown was necessary to carry the four video channels and consequently, group delay of the diplex filters required for both the forward and return transmission through the same amplifier station is at its peak at the higher end of the sub-low spectrum. Also the V.S.W.R. and roll-off in a cascaded system degrades rapidly at the lower end of the spectrum.

The point of origination of all these video channels are located at the studio where the video information whether it is on 3/4" tapes from V.T.R. machine, live programs directly fed from the camera, M.S.I. or off-air demodulated signals are then modulated in the sub-low frequency of T7, T8, T9 and T10.

3.2 Amplification:

Due to the complicity of the situation, where we need four video channels to be transmitted over 6 miles in distance, and because of the layout of the return system is usually governed by the forward system design, specifically based on the spacing between the forward amplifiers, we have elected to utilize high performance, low to medium gain reverse amplifiers. These amplifiers use the latest state of the art push-pull circitry available. These amplifier modules were installed in the same housing as the forward amplifier modules.

Except for the occasional situation in which losses are minimal between the amplifier housings, a reverse module was required at each forward amplifier location.

Extensive studies have been made to align these amplifiers to carry the four video channels with the least effect on group delay and system roll-off because of the Cascaded number of the diplex filters.

3.3 System Length:

Bi-directional two-way cable TV system requires additional design consideration over and above that required for normal down stream (forward) cable distribution system.

The parameters which governs the return path Cascade length are somewhat different from those governing the forward path. In a forward system it is mainly the limits imposed by noise and overload level degradation. In a return system, however, several factors affect the system size:

a) Noise: In the return system, noise is additive in travelling towards the head end. If the carrier to noise ratio is specified at the headend, then noise consideration places a constraint on the total number of amplifiers in the system. This means that in a system with several trunks, one or more of these individual units may be adjusted in length to allow the total number of amplifiers in the system to be within the limits set by the carrier to noise ratio consideration.

b) Overload level: In a forward system, a conventional "V" diagram based on amplifier noise output and overload levels, established the maximum Cascade length. In the return system however, because of the noise constraints discussed above, the technique can only be used to set a theoretical upper limit on the length of a single trunk run.

c) Group Delay: The diplex filters used to separate forward and return signals in the two way trunk stations produce an added time delay which is referred to as "differential group delay" or "chroma delay" between the colour sub-carrier and the video carrier in channels near the edge of the filter pass band.

The major sources of chroma delay in a sub-split system are as follows:

CHANNEL	SOURCE	CHANNEL	SOURCE
T7 and T8	Line splitter	T9 and T10	Cable Equalizers
	Directional coupler		Diplex Filters
	Directional taps	2,3, and 4	Diplex Filters
	Intermediate Bridger		
	Station Mother Boards		

Whether a designer defines his chroma delay limit as 100 nano second from the Department of Communications, Broadcast procedure 23 or as commonly accepted to be 200 nano second. This limit can be very readily exceeded with a sub-split system if the Cascade length is overlooked.

In the Scarboro system the number of forward stations between the studio and the head end is 16 stations out of these are 8 reverse amplifier modules, 4 A.G.C. automatic gain control, and 4 M.G.A. manual gain control.

3.4 Signal Ingress:

Radio frequency interference (RFI) is another design parameter that requires special consideration for the two way cable system. The upstream cable system can be faced with a hostile R.F.I. environment since many different frequency allotments and services are provided in the 10 to 30 MHz band. The return system operates in a very crowded portion of the frequency here relatively high power radio frequency sources (short wave broadcast, amateur and citizen band proliferate). These in turn, can cause contaminating signal ingress into the return system through drop cable, loose fittings, etc.

The leakage into the system through loose fittings or bad joints is a construction problem, and is dealt with as such. It can be avoided by careful installation and the elimination of poor shielding condition.

The connectors used in two way cable system should be chosen carefully with attention given to the technique used to ground the connector to the cable sheath.

In the Scarboro system we are eliminating the ingress problem by:

a) choosing active and passive equipment with adequate R.F. shielding as specified through BP 23, 24.

b) choosing connector with adequate shielding.

c) choosing double shielded or solid sheath crop cable.

and beside that we are patrolling the system for radiation detection routing to spot any area where ingress can be encountered.

3.5 Level Control:

The desired signal level can be controlled at amplifier locations by:

a) Temperature level control (TLC). This method relies on an exposed thermistor which stays at the same temperature as the cable and increases or decreases the amplifiers output.

b) Pilot Carrier automatic gain control (AGC), where a pilot has to be inserted into the return band.

c) Associated automatic gain control (AGC), which means the AGC voltage would be derived from the forward direction.

d) Composite automatic gain control (AGC) and we use this system in Scarboro as it proved to be adequate.

The cable attenuation changes are pre amplifier are less than 1 dB at 30 MHz for temperature range of $-40^{\circ}F$ to $+ 140^{\circ}F$. This is why we have elected to use AGC composite type and we are using one AGC module in each second reverse station.

IV. Making The System Work:

In the preceding paragraphs, we have made several key points in the characteristics and design aspects of a two way system. These points, in fact, should apply to both sub and mid split two way system.

The usage of the sub low VHF to carry four channels from the studio to the head end over a 6 mile distance was performed satisfactory with a minimum chroma delay, noise and distortion addition. Once received at the head end these four channels were processed to the VHF spectrum through a signal channel processor.

T8 (11.75 - 17.75 MHz) is processed to channel 10 (192 - 198 MHz)
T7 (5.75 - 11.75 MHz) is processed to channel M-1 MHz (233 - 239 MHz)
T9 (17.75 - 23.75 MHz) is processed to channel N-1 MHz (239 - 245 MHz)
T10(23.75 - 29.75 MHz) is processed to channel O-1 MHz (245 - 251 MHz)

These channels are then combined to the system at the head end from there distributed all over the system.

We have been carrying these channels using the two way sub split system over the last two years, with success. In fact, these channels are some of the many factors upon which the cable signal is sold to our subscribers.

Another experiment which will be added soon to Scarboro system, also using the two way sub split feature is a scanner transmitter receiver system which would allow viewer polling automatically.

The ability to determine which households are watching which television channel at any given time is extremely important. Present survey techniques are either very approximate and therefore unreliable or are extremely expensive. Advertisers have a very major interest in knowing which programs, and if possible, which commercials are being watched. The programmers themselves would very much like accurate mesaurements of their audience. Ultimately the best way of measuring viewership for Pay TV might be through some form of regular set "polling" rather than through an expensive terminal scrambling device. It is particularly important for the experiment that we be able to measure viewership of programs in order to assess the impact of the extra channels being offered to the viewers. Only in this way will it be possible to obtain an accurate assessment of the usefulness of the educational programming in the experiment described above. With such measurements it would also be possible to determine the impact of carrying increasing numbers of off-air and closed circuit channels.

The system comprises a scanner located at the CATV head end and the two way terminals located in each subscribers' home.

This system will monitor through the two way converter installed at the subscriber's house and the channel selected.

The scanner periodically (or via manual activation) sends out a coded number for each terminal in numerical sequence until it reaches the final number for which it is programmed. The scanner waits for a reply from the interrogated terminal before proceding with the next number.

The scanner prints out the selected channel number along with the terminal number day and hour. Terminals are addressed in sequence at a rate determined by the printer response.

The system was designed to operate in the frequency allocations specified as follows:

a) Forward: No Scan Frequency   73 MHz

                "0" Bit Frequency   74 MHz

                "I" Bit Frequency   75 MHz

b) Reverse: "0" Bit Frequency   27 MHz

                "I" Bit Frequency   26 MHz

Each transmitter operates for 256 micro seconds after the receiver terminal is interrogated, and hence, this interrogation can take place during the simultaneous transmission of the TV signal without any perceptible interference to the picture.

Starting soon this system will be tested in a chosen area of the Scarboro Cable TV system. Initial testing will be performed with 30 prototype terminals offered to the same number of households, where practical within this test area.

Conclusions:

In review, it appears the usage of the sub split, bi-direction system is quite workable and is presently the least expensive of all two-way methods. Proper emphasis should be put on the upstream portion of the system, noise built up, RFI leakage and proper equipment used in the studio, head end and the field.

With regard to the scanner project, the studies and results to date have corroborated and considerably extended previous test evidence indicating basic soundness of this two way system design.

With regard to public acceptance, the tests will constitute a serious and significant attempt to interact with the life style of the public to determine the role best suited to this new technological medium and to demonstrate and measure its public value and marketability.

The programming seasons at Scarboro Cable TV/FM have become more exciting and offer more involvement with each new year. 6,000 people have been guests on our various programs or have been involved in mobile production. Behind the scenes, ten staff members and over ninety community resource personnel produce, direct and script thousands of programs during one season.

We think Scarboro Cable TV/FM is unique. From our studios we originate 4 local channels with 24 hours of programming and digital information on all. To capsulize, we are involved with, Children's programming, Entertainment, Religion, Political Coverage and Specials.

CHILDREN'S PROGRAMMING: We believe in children being on camera and producing and directing the flow of the program. Entire schools have worked on television productions such as "News for and By Kids"; a special on Medieval Castles, Magic, Super 8 film Documentary and stories written and read by children. Our most success-ful and popular program by kids is "Kids Can Cook". A nine year old hostess Michelle teaches children how to cook and become acquainted with the kitchen. Her guests include both children and adults such as our Mayor and the Lieutenant-Governor of Ontario.

ENTERTAINMENT: Our mobile camera goes on location to cover programs featuring Ukranian Dancing, singing, and many unique neighbourhood events have made their way into the homes of our subscribers. Our most entertaining weekly series is one entitled "Quiz". This program fulfilled all needs that a good series must have. It entertained, educated and most importantly involved the viewer. "Quiz" is a half hour live phone-in program that invites the viewer to quiz the hosts and other viewers. The telephone response is tremendous and letters appear by the hundreds every week. "Quiz" did not end when you went off the air, it continued around many kitchen tables and even involved grandparents.

RELIGION: We have opened our studio facilities to all religious groups. "Twenty-nine Churches" is a weekly half hour program. By using this non-selective method we not only please all the Church groups but also provide non-biased religious programs to our subscribers. We also cover seminars and studies that tend to explore the direction of religion today.

POLITICAL: Scarboro Cable is honoured by the response from the municipal, provincial, and federal politicians. Not too many TV stations have regular taping sessions with political members from all political arenas. The viewer is given the opportunity to see and hear his elected member first-hand and participate in any discussion via live telephone programs.

SPECIALS: We try to keep our series programs at a minimal level to allow air time for special productions. Specials have included such extravanganzas as two Telethons for Multiple Sclerosis, which donated a total of $27,000.00 to research. For eight consecutive days our viewers had their first look at Sports for the Disabled when we brought to their home live coverage of the 1976 "Olympiad for the Physically Disabled". Amputees lifted 500 lb. weights. Wheelchair basketball teams from all over the world competed. The blind run. The legless swim and armless dive. Eight days of sport and sportsmanship on our twelve Toronto cable systems -- live.

Our main objective at Scarboro Cable is a concern regarding the type of programming that is sent over the air-waves and its effect on our children. We felt so strongly about this that in October we held 4 specials on "The Positive Use of Television". Educators, television producers, children and parents culminated their thoughts and discussed them on four one-hour open line shows. These programs then led to Children's Television Day in Scarborough. At our City Hall we held a one-day event on children's television. Parents and their children had an opportunity to see how a TV show is produced as well as having an opportunity to discuss programming with those behind and in front of the cameras.

TEACHING BY CABLE: THE EXPERIENCE OF ONE UNIVERSITY

Education is a natural use of Cable TV. One experiment about which you may have
heard, was written up by Ralph Lee Smith, author of the Wired Nation in August 1975
issue of "Planning for Higher Education".

I quote:

Anyone who thinks that the use of technology in education tends to deaden individuality
or foster conformity would be hard put to explain some of the things that are happening
at Oregon State University. These and other notions have undergone some interesting
tests during the university's 10 year old program for using cable television to
deliver course material. Originally launched to meet certain specific needs, the
program has generated information and experience that bear on problems confronting
educational planners and administrators today.

The Oregon State University experience also bears on one of the biggest puzzles in
the emerging world of American communications - the relationship between Cable TV
and education. Cable, with its many channels, its low transmission cost, and its
exact system of local distribution, appears to offer major opportunities of learning.
But, despite a number of exceptions, the gap between promise and performance has so
far been wide. Many colleges and universities, although intrigued by cable, have
been baffled about how to organize and implement its effective use.

Education could receive low priority in policy making and system planning in urban
centers. Access under favorable economic arrangements, and substantial channel
capacity for educational delivery, could both be lost. At a time when educational
planners are seeking the twin goals of outreach and reduced cost, this loss is one
that higher education cannot afford.

One of the most extensive of these is at Oregon State. Located in Corvallis, a city
of 30,000 Oregon State University is a land grant institution whose enrollment has
been limited by the legislature to 15,000 students. Since 1966 it has transmitted
much of the content of many courses over the local cable system, Corvallis Cable
Company. This sytem is linked by microwave to systems in two nearby towns, Albany
and Lebanon.

The videotaped programs are transmitted from the Classroom Television Center on the
OSU campus, which has a direct cable link to the Corvallis Cable Company headend.
From there, the programs go out over a channel on the cable system that is reserved
for university use. They are simultaneously shown on the university's own closed-
circuit system, which links 18 classroom and dormitory buildings, so that students
can view them at convenient campus locations. Enrollment for the courses totals
8,600 per year - over half the total enrollment of the university. Seventy percent
of the viewing by students is done off campus. The cable system has about 10,000
subscribers.

From "Failure", the Seeds of Success

Oregon State's experience with TV goes back to 1957, and originally did not involve cable. In that year The Ford Foundation funded an experiment for intercampus exchange of courses by TV among Oregon State, the University of Oregon, the Oregon College of Education, Portland State College, and Willamette College. The state's first public television station, KOAC, was built with Ford funding to transmit the programming to all the participants, when it was found that transmission by telephone company facilities would be prohibitively costly. Courses were created at each institution which could be taken by TV, for credit at the others.

The exchange program ran until 1963 when it was phased out. By the most generous estimates it was not an over-all success. Professors teaching a given subject at one institution were often less than enthusiastic about having students enroll for imported TV courses with similar content. Ironically, the higher the quality of the import the cooler its reception by some resident professors. By the end of the experiment there was little enrollment for any of the courses on campuses other than on which the program originated.

Certain professors developed a positive interest in TV's potential as a teaching tool. A significant lesson was buried in the debris of the old program. One course did notably well. It was an unpretentious one-credit offering in General Hygiene, which maintained strong inter-university enrollment while more ambitious courses languished. The reasons were simple. Several of the participating institutions wanted to offer the course, but only one had a full-time faculty member with experience in teaching it. It filled a definite need, and filled it economically.

In 1960, OSU had an immense increase in student enrollment. Between 500 and 1,000 students were signing up quarterly. Such enrollments swamped the capacity of the departments for teaching it traditionally. The problem was compounded by the fact that many of the heavy enrollment courses were sequenced, so that students could not take the second or third segments until they had taken and passed the first. Students' progress through the university was threatened by this blockage.

The university faced two alternatives; to conduct auditorium-sized classes, or to hire many new faculty members and embark on a building program to create more classrooms. The second alternative was never a practical one within the limits of the budget.

Coincidentally, in 1964 the town of Corvallis granted a franchise for the building of a cable system. The university saw in this a possible solution to its problem and approached the cable company with a proposal of potential mutual benefit. The company was receptive and an agreement was reached which provided that one of the system's 12 channels, channel 5, would be assigned to the university. The cable company also agreed to donate the labour for the wiring of a closed-circuit system on campus, with the university providing $6,000 for the materials.

On September 1, 1966, Oregon State University went on the air, or, more accurately, on the cable, with a program of course offerings that has since become one of the most extensive, and probably one of the most successful, in the country.

Private corporations, in business for profit, are not noted for giveaways. What, then, was in it for the cable company? As it happens, its owners had a strong personal interest in communications technology for education. But more to the point, they had a practical incentive, too. Through the arrangement with OSU, the company receives a large amount of programming free of charge, which can and does increase the number of its subscribers.

It is now nearly impossible to rent a room or apartment to a student in Corvallis if it does not have Cable TV.

The company attributes 30% of its subscribership to the programming that it receives from the university - which, at a monthly subscriber's fee of $5.50, amounts to almost $200,000 per year.

Making and Using the Tapes

In one pattern followed with variations by a number of departments, students enrolled in a three-credit cable course watch two videotaped lectures per week made by the professor giving the course, and then attend a one-hour recitation period conducted by a graduate assistant. Students may have individual contact with the professor during his office hours. Four-credit courses include three taped lectures per week plus the recitation; or, in the case of science courses, two taped lectures plus a recitation period and a lab.

A professor wishing to go on the cable with his course meets with the center's director and production manager for an orientation in the use of TV. He then outlines the series of lessons, bearing in mind the one ground rule established by the center - that, for each hour-long lecture, the professor cannot be on camera as a "talking head" for more than 15 minutes. The rest of the lecture must be composed of visuals for which the professor provides commentary, or visuals that have their own sound track.

As many as 200 hours of preparation may go into the making of some of the one-hour tapes. Things are easier for the professor during the next several years while the tapes are being used, but he does not drop out of the picture. Each year he updates various portions of the tape series. After about three years, most professors undertake a revision of the entire series, sometimes making a full set of new tapes.

Imagine a two way terminal in each student's office, and at home. Computer assisted education aids have been shown to be both practical and successful by many studies including the Nitre Corporations's experiments in Reston, Virginia.

The student, the institution, the cable company all working together to study the results all round the world.

As the Global village develops and knowlege of television's influence is more completely understood, great care must be taken to have the courage to say "NO" to development if the results are pretty certain to be negative.

You must examine alternatives by joint forces with the software producers. Engineers, whether Research & Development, or practical must face up to their Responsibility. You must be responsible for the delivery system and ultimately the product delivered on that system.

You cannot shirk it. You cannot leave it to Governments alone. You cannot leave it to business alone. Research results of import must be implemented..by whom? By you and me. Corporate responsibility must stretch to software and studies of television on the individual.

Without realizing, engineers are the people who set the wheels of change in motion and become the "true leaders of the community". So be up-to-date; take the initiative; and most of all be Responsible.

R   REGIONALISM

E   EXPERIMENTS in Engineering

S   STATE Government -- joint ventures with industry

P   PROFITS - only Part of the Equation

O   OWNERSHIP - A mix = Government - Common Carrier - Private

N   NEIGHBOURHOOD -- the community

S   SOCIAL RESERACH and results studied

I   IMPROVEMENT -- in the quality of life

B   BANKING

I   INITIATIVE

L   LIVING AND LOVING

I   IDEALS

T   TELECOMMUNICATIONS

Y   YOURS

"YOU ARE THE BOWS FROM WHICH YOUR CHILDREN, AS LIVING ARROWS, ARE SENT FORTH".

If we find truth in this quotation from Kahlil Gibran, we must take implicit responsibility for the future of their world. As our lives and society become more complex we must be sure to shape ourselves as symbols of solid responsibility to the children in our community. Sometimes we take our freedom and our democracy for granted forgetting that we tend to live up, or down, to the expectations of others rather than our own.

Let us try to shoot these arrows forward, high into the air to the horizon lit by the sun's rays. Let us accept our responsibility.

Zweiweg-Kabelfernsehsysteme

Eines der Themen aus dem Kabelfernsehbereich, die gegenwärtig ein sehr starkes Interesse finden, ist das der "doppelt gerichteten" Kabelnetze. An die Dienste, die sich mit der Einführung eines voll ausgebauten Zweiweg-Kabelnetzes erreichen lassen, werden eine Reihe von Forderungen gestellt. Dienste, wie z. B. Sicherung der Wohnung, Abstimmung durch die Zuhörer oder -schauer, Untersuchung der Verbrauchsgewohnheiten, örtliche Programme für bestimmte Gruppen aus verschiedenen Wohnbereichen usw., finden bereits Anwendung und können noch erweitert werden.

Der vorliegende Beitrag behandelt einige Aspekte, die sich im wesentlichen auf die Nutzung des Rückkanals in einem doppelt gerichteten Kabelnetz beziehen. Es werden die wichtigsten, beim Aufbau und beim Einsatz von Zweiweg-Kabelsystemen zu beachtenden Faktoren besprochen. Der Beitrag beschäftigt sich nicht mit dem Versuch, bestehende Verteilnetze mit Rückkanälen nachzurüsten, obwohl dafür einige der Überlegungen durchaus Gültigkeit besitzen.

Video Information System Development in Japan

T. Namekawa
Suita Osaka, Japan

ABSTRACT

An information distribution system for local community servics in which
two-way analog and digital signals are transmitted using a coaxial
cable system, has been researched and constructed in small prototype
subsystem by my cooperated research group in 1971.

That was a new hibrid information distribution system which is capable of
controlling digital signals and exchanging 40 video channels for 300
subscribers. The maximum frequency bandwidth of this system expands to
300 MHz, 24 TV channels, 40 CCTV channels and two-way digital channels
of 2.5 Mb/s are transmitted and processed in this system.

After that, more complicated and extended system named " Video Information
System " has been planned and developed as a Japanese national pilot
project from 1972. This is a new home life information system using a
limited two-way communication system and computers. They have constructed a prototype in a show room at Higashi-Ikoma, nearby Nara, using
a new optical fiber communication system instead of coaxial cable system.

I will describe the detail of our old prototype subsystem and a short
history of developement of the several numbers of pilot projects in
Japan.

The paper concludes that the problems concerning what would be and how
to get the best broad-band cable information system is not solved, and
we need more continued experimentations in different communities.

1. THE PROTOTYPE TWO-WAY DISTRIBUTION SYSTEM IN THE EARLY TIME (1)(2)

A joint research group organized by Osaka University, KEC, and other four manufactures in Kansai area has proposed and developed the new information distribution system for local community which includes two-way digital and two-way video networks to realize the new idea. We have completed the model system experimentally.

The maximum frequency bandwidth of this system expand to 300MHz. 24 TV channels, 40 CCTV channels and two-way digital channels of 2.5Mb/s are transmitted and processed in this system.

Signals used for several services to transmit and distribute using the only one coaxial cable are divided into the following types:

(a) Two-way digital signals for
  (1) Tele-controlling
  (2) Automatic inspection of meters
  (3) Data transmission
  (4) Automatic alarm
(b) Two-way analog signals (CCTV) for
  (1) TV telephone
  (2) Facsimile
(c) One-way analog signals for
  (1) Retransmission of TVs
  (2) Retransmission of FMs
  (3) Local originated TVs

With above classifications we have contemplated the frequency allocation of coaxial cable and decided the allocation as shown in Fig. 1.

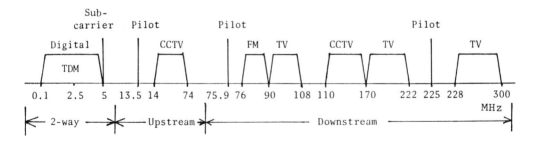

Fig. 1. Frequency allocation of the system.

Fig. 2 shows the blockdiagram of our system. Information center (IC) controls the various kinds of information sources. Sub-center (SC) connected to IC by multiple controllers 2.5Mb/s digital signals and CCTV channel exchanged by the subscribers. Relay network (RN) is capable of transmitting and distributing two-way signals using the only one coaxial cable. Subscriber terminal (ST) which we call Automatic Information Retrieval Controller (AIRC) controls the various kinds of information signals for each subscriber.

In this system one sub-center is capable of controlling 300 subscribers. However in our experiment we have made only one sub-center, relay network two subscriber terminal and the terminals for sub-center and two homes, which is sufficient to obtain the various datum.

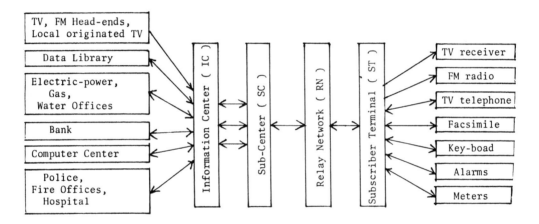

Fig. 2.   Block diagram.

1-1   DIGITAL PORTION OF THE SYSTEM

One of the outstanding features of digital communication system is that it has made possible to reduce a various kinds of many interesting properties by performing a certain logical operation on an input information symbol sequence. These sequence are

(1) Error detecting or correcting code
(2) DC restricted code
(3) Synchronizable code

Besides these a digital signal sequence can be used for controlling and switching the various analog signals as well as digital signals. Our system enjoys these advantages of using digital signals.

Our system transmits the digital signals together with the various analog signals using the same coaxial cable, i.e. in a hybrid form. The reason is that the hybrid system is more suitable when we use digital signal sequence for controlling or switching the various analog signals. As shown in Fig. 1, the spectrum of digital signals occupies the lower frequency band, where the transmission loss is relatively low. As almost all subscriber's terminals in future CATV system are predicted within a mile from the sub-center, digital signals will be transmitted without a digital signal repeater.

The framing scheme is shown in Fig. 3. The framing pulses are altern-ating 1, 0, 1, 0, ···· pulses. Upstream frame starts from the framing pulse " 1 " and downstream frame, " 0 ". The framing frequency is 840Hz. Downstream digital signals are multiplexed at the sub-center in ordinary way. The framing circuit at the subscriber's terminal demultiplexes its channel by counting the number of pulses from the framing pulse " 0 ". Digital signals from each subscriber are multiplexed in somewhat diff-erent way compared with the ordinary PCM systems.

The reason is that each subscriber's terminal is located in geographi-cally different place. The propagation time required for digital signals depends on the distance between each subscriber and sub-center. To compensate these differences existing in the propagation times, our system uses two circuits, i.e., a circuit which adjusts roughly the propagation times and a circuit which adjusts precisely the remaining difference. Since it is difficult to compensate completely the diffe-rences, each channel in downstream frame has a guard space of 1 time slot as shown in Fig. 3.

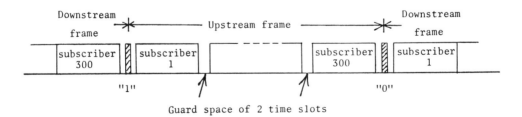

Fig. 3.   Line format of digital signals.

Among the many significant problems we encounter in CATV systems, keeping each subscriber's privacy would be perhaps one of the most important problems. For keeping each subscriber's privacy, our system exploit the error detection abilities of digital signals. Among the digital signals, especially the control signals of various kinds should be kept highly reliable. Since our digital channels are two-way, we adopt ARQ (Automatic Retransmission Request) system for improving the reliability. Error control schemes for our system are summarized as follows.

(1) ISO code of 7 bits is encoded into a code of minimum distance 2.
(2) Among the digital signals, especially important signals such as address pulses,control signals etc. are forced to retransmit and check the coincidence. The error control circuit continues to sends " NAK " until two successive block are coincident.

Thus the probability of the occurrence of undetected errors for the valuable digital signals would be surprizingly small even if the channel error rate increases to 10^{-4}.

1-2 ANALOG PORTION OF THE SYSTEM

The analog signals handled in this system are one-way TV and FM signals and two-way CCTV signals.

The retransmissions of FM signals and 3 TV signals are set in 76-90MHz 90-108MHz, respectively. 9 TV channels are set in 170-222MHz according to the ordinary Japanese broadcast frequency allocation as shown in Fig. 1. These are similar to a standard CATV system.

Local originated TV programs are transmitted in 228-300MHz divided by the 12 channels.

Two-way CCTV signals are divided into upstream and downstream by FDM basis. Upstream signals are sent in 14-74MHz and downstream signals in 110-170MHz.

Each CCTV channel is transmitted by 1.0MHz VSB signal and the guard band is set to 0.5MHz both in upstream and downsteam. In case that 300 subscribers use these 40 channels, it would be most desirable to access the channel randomly. However considering the implemetation of the system, we have rather adopted the simpler scheme for channel selection. That is, as shown in Fig. 4, 2 channels are assigned for 16 subscribers. In addition, considering the mutual interference of adjacent channels and the simplification of apparatus, adjacent two channels are assigned to one subscriber with a guard band of exactly equal to one channel. Therefore the necessary bandwidth of RF tuning circuit is 4.5MHz, so that all the circuit configuration are the same as a standard TV receiver.

A new type of electronic exchange system at sub-center is devised for the system simplification. The problem of the system simplification is urgent especially in local community system.

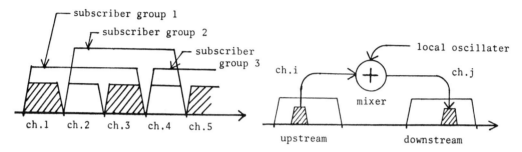

Fig. 4. CCTV channel assignment for subscribers.

Fig. 5. Simplificated video exchager.

One of the ideas is shown in Fig. 5. Some of the modulated signal in upstream is directly converted to another subscribers in downstream by frequency conversion at the sub-center. As far as video signal of 1MHz is concerned. This is the simpler method for video signal of 1MHz since this does not require to bring it back to base band.

While CCTV channels are used for TV telephone of 1MHz band width, they are also used for facsimile channel by the electronic control with digital signal. Facsimile signal is also taken to the VSB signal of 1MHz bandwidth.

The performance-monitoring for analog signals is done by the pilot signals located at 75.9MHz and 225MHz in downstream and 13.5MHz in upstream respectively.

2. SYSTEM DEVELOPMENTS GO

CATV systems originated from the needs for the TV signals retransmission services. However, the cable's capability to perform a broad band communication with backward channel functions was widely recognized at the end of the 1960's.

We have developed the Two-way Information Distribution Subsystem for local community which includes the two-way high speed data communication and the limited two-way video communication using only one coaxial cable. (1) (2)

We succeeded to operated the model, however, this experimentation offered only a technical capability of video information service for 300 local subscribers.

There appeared among those in the government and private circles an active move to forward the development of a new information system for local community.

A research committee was organized under the chairman Prof. Y. Hayashi by the aids of Ministry of International Trade and Industry. The committee, named "Local Community Information System Research Committee" worked on the field of research and survey concerning the potential needs, capable technology, status of industry and relating regulations in 1971. They continued their reseach work doing some case study up to 1973. They could not get complete consensus on the basic concept of local community and its functions relating to community communication.

At the end of 1971, they suggested to promote the national project constucting a model Information systems at new towns. They expected to make clear the real needs for some kinds of local information system supposed to improve the community function and the way of daily life for the individuals, by the test and evaluation on the pilot system. They expected also to make clear the technical and economical capability, when the broad-band distribution networks with backward channel is applied to the information services for the improvement of community function and daily life.

The " Video (or Visual) Information System Development Association " - VISDA - was established on may, 1972, under license of the Ministry of International Trade and Industry.

Video Information System is a two-way information system enabling
collection, distribution, processing, and display of multipurpose
information through a combination of coaxial cables (now optical
fibers) and computers. (3) (4)
The general requirements were as follows.

(1) voluntary selection and reception of desired information.
(2) continuing education.
(3) promotion of computerizatin in our daily life.
(4) rearing of video industry.

Video Information System is a large scale social system and its development must be attempted by setting an actual field where the system
is activated. They selected and decided the site of the model-town in
Ikoma city, Higashi-Ikoma area. This site is situated at the south
part of old city of Nara. They expect 300 volunteers to join this
information system as subscribers in this area.
The following aspect is one part of the targets of this model system
development.

(1) To discover technical feasibility of Video Information System
through new media and the development of new services.
(2) To develope and provide new software, suitable for Video Information
System.
(3) To take the public's understanding and awareness concerning Video
Information System.
(4) To make clear the potential needs and to create new needs by the
research survey in this model town.
(5) To examine economics of business operation, with full consideration
of the service content and the demand.
(6) To make social assessment.

Development Program designed at the beginning is as follows.
The system development program, designed to achieve the above mentioned
targets, was being carried out over the 7 year period from 1972 to 1978,
But, we have experienced the bad state in our national economy by the
oil shock, and they have had to decide to delay their time schedule
more two years later.

(1) First year (1972)
Established the Video Information System Development Association, a
juridical foundation.
Made a basic design of the system, and invited the public to offer a
model-town site.

(2) Second year (1973)

Made a detailed design of a total system.

(3) Third year (1974)

Made detailed design of sub systems and program software.

(4) Fourth year (1975)

An exhibition room was opened at the headquarter of the Association in the midst of Tokyo, and they launched a small pilot run of Video Information System equipment and software.

(5) Fifth year (1976)

They checked again their system design and corrected it in some part. Most significant change was the adaptation of new optical communication system using optical fibres instead of usual coaxial cables. We have no natural resources such as like as the copper, and also the copper mine is becoming poor now on the earth surface. I think it is very imtortant movement to utilize the glass fibre cable as a new medium of communication, especially in the field of CATV.

An show room prototype system was opened at the Higashi-Ikoma site on September and opened to the people of the local community.

Phase 1 was completed here. Phase 2 started and they called it by the name " Higashi Ikoma Optical Visual Information System " - (Hi-OVIS) -. The detail on this Hi-OVIS development is described by aother paper of this symposium presented by Dr. M. Kawahata. (5)

The history of practical CATV tells us the following three generation.

  1st generation ; High quality, More channels
  2nd generation ; Local origination
  3rd generation ; Two way services

" Tama Living Information System Development Association " was established under license of the Ministry of Post and Telecommunication in 1972.

The system, called by the name CCIS - Coaxial Cable Information System - aimed at the project to construct pilot system and to evaluate it from economy, management, payability and technology. Their experimental system was constructed and operating now from Jan. 1976 at their site Tama new town.

Another paper on this symposium titled " Coaxil Cable Information System with Interactive Television Services " presented by Mr. K. Yamaguchi and S. Yoshida describes the system and its operation on services. (6)

From 1973, the two Association mentioned above got together as two Divisions of one organization named Living Visual Information System Development Association.

Nippon Telegraph and Telephone Public corporation (NTT) has continued their experiment in the field of two-way CATV. They succeeded in their field trial of a new two-way broad band CATV system applied for individual still-picture communication. (7) New prototype system that is a video information system for IR and CAI using telephone pair cables and home TV receivers, is now constructed. The detail is described by the symposium paper presented by Mr. H. Nakajima. (8)

" Rural Multi Purpose Information System Association " was established under the license of Ministry of Agriculture and Forestry in 1975. They are working their research on the new communication medium called Multi Purpose Information System - MPIS - for the rural community.

I have mentioned above the names of main pilot projects in Japan. There are other several numbers of development projects on two-way broad-band CATV systems at the present.

3.CONCLUSION

I have described our experimental study on the two-way information distribution system for local community, and also talked about a short history of development plojects on the field of broad-band distribution networks with backward channel in Japan.

So much work has been done and is being done, however, we have unsolved problems such as where we could find out true demand, how to take advantage of the optical fibre communication technology. I am convinced we have to continue various kind of experimentations at different type of communities.

REFERENCES

1. Namekawa, T.: Local Community Information System. JIEE Japan vol.92 no.11, 79-84, 1973

2. Namekawa, T., Kasahara, M. and Murata, M.: Two-way Information System for Local Community. Wescon Technical Papers, IEEE, session 284, 1972

3. Namekawa, T.: Comutor application on CATV system. JITE Japan vol.28 no.4, 327-331, 1974

4. Shimo, K.: On the Development of the Video Information System. JIEE Japan vol.93 no.1, 33-37, 1974

5. Kawahata, M.: Hi-OVIS Development Project. Symposium, Munchner Kreis, April, 1977

6. Yamaguchi, K. and Yoshida, S.: Coaxial Cable Information System with Interactive Television Services. Symposium, Munchner Kreis, April, 1977

7. Maeda, K.: Individualized Still-Picture Communication on a Two-way Broad-Band CATV System. IEEE Trans. Commun., vol. Com-23 no.1, 104-107, 1975

8. Nakajima, H.:Video Information System for IR and CAI Using Telephone Pair Cables and Home TV sets. Symposium, Munchner Kreis, April, 1977

Die Entwicklung von Fernseh-Informationssystemen in Japan

Ein örtliches Informationsverteilungssystem, bei dem analoge und digitale
Signale in beide Richtungen über Koaxialkabel übertragen werden, wurde 1971
von einer Arbeitsgruppe entwickelt und ein Teil des Systems als Muster gebaut.

Mit diesem neuen Hybridsystem konnten gleichzeitig digitale Signale übertragen
und 300 Teilnehmer über 40 Fernsehkanäle versorgt werden. Die Frequenzband-
breite dieses Systems beträgt maximal 300 MHz. Es stehen 24 Fernsehkanäle,
40 Closed-Circuit-Fernsehkanäle und Datenkanäle mit einer Übertragungsgeschwindigkeit von 2,5 Mbit/sec. in beiden Richtungen zur Verfügung.

Nach 1972 wurde das komplizierte und größere "Video-Informations-System"
konzipiert und als japanisches Pilotprojekt entwickelt. Dieses neue Informationssystem für den privaten Bereich ist für den begrenzten Dialogverkehr mittels
Rechner ausgelegt. Eine Versuchsanlage, bei der anstelle von Koaxialkabel Kabel
aus optischen Fasern eingesetzt werden, steht in einem Ausstellungsraum in
Higashi-Ikoma bei Nara.

Im vorliegenden Beitrag werden Einzelheiten des alten Mustersystems beschrieben
und eine kurze Entwicklungsgeschichte der neuen Pilotprojekte in Japan gegeben.

Der Beitrag kommt zu dem Schluß, daß eine Lösung der Probleme, wie ein
optimales Breitband-Kabelsystem beschaffen sein muß und wie dies realisiert
werden kann, noch aussteht und dazu noch weitere Versuche in anderen Stand-
orten erforderlich sind.

Coaxial Cable Information System With Interactive Television Services

Kofumi Yamaguchi and S. Yoshida
Nagayama, Tama-shi, Tokyo, Japan

Abstract

After studying the possibilities of cable television as a community information
system for one and half years, the research panel in the Ministry of Posts and Telecommunications decided to conduct an experiment on this system in an actual community.

Consequently, a Coaxial Cable Information System, CCIS which includes interactive
television services was established in the suburbs of Tokyo and the operational
experiment has been under way since January 1976.

Generally speaking, the participants are kindly disposed to this experiment and
it will continue till this September. Although its result will be published after
its completion, we are reporting here the services and hardware of this system.

1. Introduction

With the complications of the economy and society and with the higher standard
of civilization, it is expected that relevant information should be provided to meet
various requirements in many fields of individual life. It is desirable to develop
diversified new information systems not only with one-way but also with two-way
communications, if such requirements are to be met. However, apart from the arguments
on the theoretical and technical possibilities, the reality of cable television,
CATV, and two-way information service, is that CATV is used only for re-transmission
of off-the-air TV signals and that the two-way information service remains in the
experimental stage.

This being the reality, there has been a movement both in the Government and
in the private sector to develop a new information system.

Two aspects in the development of this new information system require special
attention.

First, to grasp correctly the need for pragmatic information, to set up an
economical system, and to examine efficient management and payability.

Second, to develop a two-way information system with relevant video equipment
in which each household can be connected to an information center through a broadband

cable. Like two wheels of a car, both are indispensable factors for developing a good system.

Under these circumstances, the Living Visual Information System Development Association was established in 1972 under the auspices of the Ministry of International Trade and Industry and the Ministry of Posts and Telecommunications. In addition, two experimental projects on information systems were launched in Nara and in Tokyo.

This is a report concerning the experiment now under way in Tama New Town in Tokyo, which is on the Coaxial Cable Information System, CCIS.

2. Outline of the Tama CCIS Experimental Project

CCIS is an abbreviation for Coaxial Cable Information System and can be defined as an information system with a tree-like structured coaxial cable distribution network. The system will be used principally in ordinary households, aiming at information services such as original TV programme service and non-broadcasting type services as well as retransmission of off-the-air TV signals. It is a technical term used for an advanced stage of CATV. The coaxial cable for CATV, which is used mainly for eliminating shadow areas in over-the-air broadcast TV signals, has the capacity to transmit TV programmes for more than 20 channels, but in fact only a few channels are used for the re-transmission of off-the-air TV signals; the bulk of the transmission capacity remains unused. Therefore, the Ministry of Posts and Telecommunications has set up a project for CCIS which would provide a variety of information for a community by making use of the transmission capacity of the coaxial cable for CATV, and has already made a comprehensive investigation.

Nevertheless, as the points at issue could not be clarified by studying a desk plan or a demand survey, it was decided that an experimental examination was to be made in Tama New Town. There, CCIS facilities were established in a real community and made use of in various ways so that a grasp of the needs of the people and an assessment of the future possibilities of CCIS could be made.

Tama New Town is located 30 kilometers west of the central area of Tokyo, and since 1965 development of this residential city has been progressing. When completed, it will spread 14 kilometers from east to west and 4 kilometers from north to south, covering 3,000 hectares of land. The high-rise apartment complex is expected to accomodate about 330,000 people in 90,000 households.

CATV (or CCIS) can be divided in its development into three generations;

first generation: re-transmission of off-the-air TV signals only

second generation: re-transmission of off-the-air TV signals and original TV programme service by introducing a studio system for local origination to the first generation CATV

third generation: re-transmission of off-the-air TV signals, original TV programme service, and other information services by adding various communication apparatuses with the two-way communication function to the second generation CATV

The second and third generation CCIS are future information systems based upon the needs of the local community. The experiment in Tama New Town is centered upon the second and third generation CCIS and has been carried out as a joint experiment of the Ministry of Posts and Telecommunications, Japan Telegraph and Telephone Public Corporation, and the Living Visual Information System Development Association.

Our experimental project in Tama New Town has five objectives:

First of all, we would like to grasp, through CCIS services, the needs of the people for an information service concerning their daily lives. Second, we seek the feasibility of commercializing CCIS services as an information system on daily life. The third objective is to study the role of CCIS in a community. Fourth is a technological verification of the whole system. The last is to arouse public interest in CCIS.

3. Services and Hardware of Experimental CCIS

There are at present about 18,000 households in Tama New Town and an experiment involving CCIS is being carried out in a small part of it. The cables have been set up in 904 households, but about 500 monitoring households are participating in the experiment.

As shown in Figures 1 and 2, Tama CCIS is composed of a reception antenna site, an experiment center, monitoring households, and coaxial cables which link together the antenna site, the center, and the households. The antenna site is 2.4 kilometers away from the experiment center. The antenna receives regular VHF-TV broadcast waves and a UHF fascimile radio wave sent from a newspaper company.

We have an office, a studio, and an apparatus room in the experiment center, where signals for various kinds of services are produced. There are now ten different services being performed in the Tama CCIS experiment. They are broadcast TV retransmission service, original TV programme service, automatic repetition telecasting service, pay TV service, flash information service, facsimile newspaper service, memo-copy service, auxiliary TV service, broadcast and response service, and still picture request service.

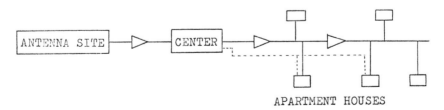

APARTMENT HOUSES

FIG.1 OUTLINE OF THE WHOLE SYSTEM

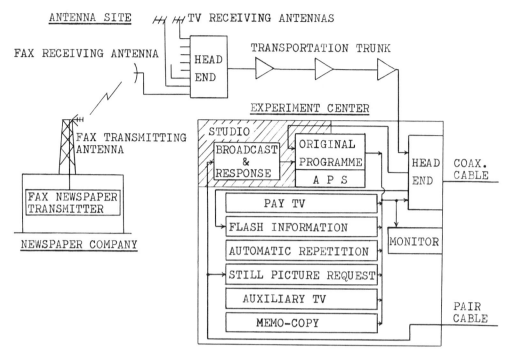

FIG.2 SIGNALS SOURCE SYSTEM OF TAMA CCIS EXPERIMENT

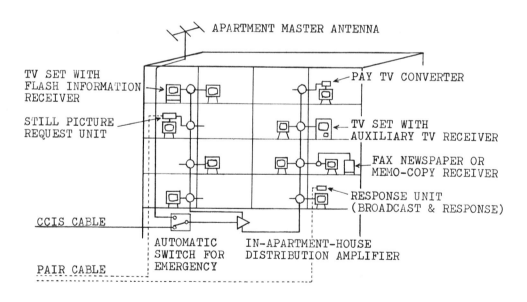

FIG.3 INTERNAL DISTRIBUTION SYSTEM OF AN APARTMENT HOUSE

The distribution system in an apartment house is shown in Figure 3.

Whether a regular TV set can be used as it is or a special terminal unit has to be installed depends upon the available services. In case something goes wrong with the CCIS transmission line, a switch in each apartment house can be used to automatically change from the CCIS cable to each master antenna in order that the whole services may not be cut off at one time. Therefore, re-transmission of off-the-air TV signals can continue at any time.

In the experiment, all TV services are in color except the auxiliary TV service. Now, the content and facilities of each service can be summarised as follows:

(1) Services without up-stream communication channels

Broadcast TV re-transmission. There are seven TV Broadcasting stations in the Tokyo area, so the number of re-transmission TV channels is also seven. As the antenna site was set up at a place where a clear picture can be received, the monitoring households can also get clear pictures.

Original TV programmes. Original TV programme service and automatic repetion telecasting service each occupies one TV channel, and a monitor can use a regular TV set.

Original TV programmes are produced in a 7 by 7 meter studio at the experiment center. Included are 3 color cameras equipped with three carnicon tubes, a telecine apparatus (for 16 mm and 8 mm film, slides, and opaque cards), an audio-video switching console, five three-quarter inch casette video tape recorders, an automatic programme switcher equipped with a mini-computer, and so on. Part of the CCIS cable route was set up for two-way transmission, by which relaying a special TV programme from a school, kindergarten or a meetinghouse was made possible.

The original TV programmes are closely related to community life, such as news and daily information involving Tama New Town.

Automatic repetition telecasting. In the automatic repetition telecasting service, up to 30 opaque cards are repeatedly telecast, one by one, by synchronizing them with audio casette tapes. Colour can be electronically added to a card with words in monochrome. This service is used for general information, shopping information, reporting by the people of the community, etc.

It is characterized by easy production of programmes.

Pay television. Pay television service uses a special frequency band not receivable by a regular TV set. So, the viewer requires a special adapter in addition to his regular TV set.

Pay television services has three TV channels with two kinds of adapters. They both are channel converters which convert signals received into the regular broadcast channel (Channel 11). However, one adapter can receive only one channel. The converter circuit works by inserting a specified ticket into the adapter.

The other adapter can receive three channels. A small plug-in-unit has to be inserted into the adapter. A key signal is transmitted from the center on a different carrier. When the signal coincides with a type of plug-in-unit, the converter is put into effect. The plug-in-unit has the function of choosing the right channel.

Pay television service is used for movies, educational films, and so on.

Flash information. Flash information service deals with five different types of information; general news, sports news, weather forecasts, community news, and daily life information.

This is visual information without sound, using only letters produced at the experiment center. The letters are superposed as digital signals on the vertical sync signal of a TV channel, which is one of the re-transmission TV signals (Channel 10).

As the number of Japanese letters is very large, letter pattern signals are to be transmitted. The signals are separated from the video signal in a viewer's terminal unit, and when he chooses his own information he can watch it on channel 10, as it appears in flash signs from right to left. There are four possible positions that this information can appear on the screen. Of course, it can also be switched off.

Facsimile newspaper. The signal of facsimile newspaper service comes from outside of the New Town. The UHF signal from a newspaper company received at the antenna site is converted into VHF and transmitted to household receivers through coaxial cables. A monitor can receive newspaper copies of actual size from his terminal receiver. The main scanning on the receiver is multi-stylus electronic scanning, and the development method is powder printing by static electricity. The scanning density is 8 lines/mm. It takes about 4 minutes and 30 seconds to receive one page of a paper.

Memo-copy. Memocopy service is also a kind of facsimile. A postcard size memo-copy is sent from the experiment center. At the center, plane scanning is made of the original information card by a self-scanning photo-diode array. A monitoring resident can get the print from his terminal unit in which a linear thermal printing head prints the memo-copy signal on a thermo-sensitive paper. There is a glazing made from two ingredients on the paper which generates colour by chemical reaction when heated. The scanning density is 4 dots/mm and the time required for transmission of postcard size is about one minute.

Memo-copy service is used for notices from local authorities and communication amongst the residents in the community, etc.

Auxiliary TV. In auxiliary TV service, a monitor has a special receiver with two screens. One is used as a colour TV to receive signals from Channels 1-12. The other is a small monochrome screen which is controlled by the signals from the center. It works automatically even if the colour TV is switched off. This service is used mainly for sending urgent information with sounds.

The colour TV has a reservation button for both original TV programme service and automatic repetition telecasting service. If the button is pressed in advance, the requested channel can be automatically received, once these services start, in spite of the fact that the TV is tuned to a different channel. The whole process is controlled by the signals sent from the experiment center.

Special terminal units have been produced for this experiment and rented free of charge to the requesting monitors. The number of terminal units produced for these services is as follows; 300 for Pay TV, 40 for Flash Information, five for Facsimile newspaper, 30 for Memo-copy, and 45 for Auxiliary TV.

(2) Services with up-stream communication channels

Broadcast and response. Broadcast and response service is an interactive television service in which viewers can respond during the TV programmes. If viewers have only a regular TV set with no special terminal unit, this service works as an original TV programme service, but if they have a special terminal unit they can send response signals. The terminal unit is a telephone type home unit with a keyboard.

The TV programmes in the broadcast and response service are English or arithmetic lessons for primary school children. A response can be given as follows; the teacher asks a question at the studio and gives five possible answers, from which a viewer can select the correct answer. When the viewer makes his selection by pushing one of the five key buttons on the terminal unit, it is transmitted to the studio. The results of the questions and answers will be indicated in the studio and the teacher can grasp the degree of the viewers' understanding of the programme. Also, a nominee of the viewers can talk to the teacher by using a telephone type microphone. The response signal can be sent through exclusive pair cables laid between the studio and the terminal unit.

As a response line we used a pair cable instead of a two-way coaxial, because for a small number of monitors as in this experiment the pair cable is more economical.

One hundred terminal units were produced for this service and about 70 are in use at present. The system for this service is illustrated in Figure 4. The signal sent by the pair cable goes into the voice reception unit and the voice of the viewer is sent to the audio system in the original TV service facilities. The response data from each viewer can be memorized and indicated, and is also sent to the central processing unit for total or percentage calculation. The result of the calculation will be indicated on the response display board, enabling the teacher and the viewers to know the results.

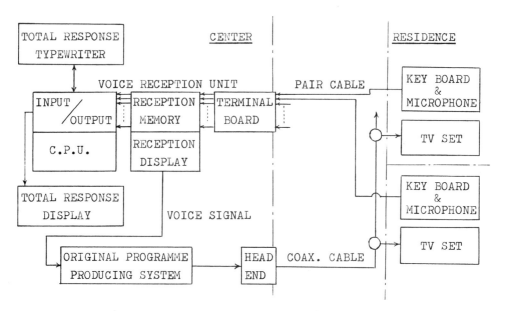

FIG.4 BROADCAST AND RESPONSE SERVICE SYSTEM

Still picture request. Still picture request service is also a kind of inter-active TV service. Still pictures which involve information in various fields, such as education, medical care, infant care, and leisure and recreation, are kept in stock (up to 6,000) in microfiches at the experiment center. A viewer can request a still picture by specifying its number with the aid of an index table. When a viewer presses the number on the keyboard of his terminal unit, the request signal is sent to the center via a pair cable.

This data is memorized by a CPU, central processing unit, at the center, as illustrated in Figure 5. The CPU gives a retrieving order to the image file. The picture taken out of the file is shot as a single frame by a colour camera. The address and data codes are attached to the picture signal and sent through a modulator to the coaxial cables.

The viewer's terminal unit has one frame magnetic disc picture memory. When the requested picture signal is received, it is recorded in the memory. The signal is read out immediately and the still picture is shown continuously on the TV screen.

As the output signal of the terminal unit is Channel 11, the viewer can see it on his regular TV set. The time interval between the request and the appearance of the picture is about 4 seconds. There are forward and backward sequence keys besides numerical ones on the keyboard. No sound is given in this service.

Thirty terminal units were produced for this project.

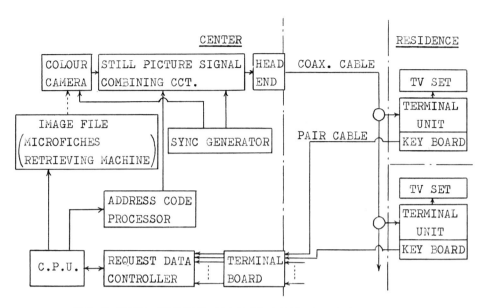

FIG.5 STILL PICTURE REQUEST SERVICE SYSTEM

(3) Transmission line characteristics

The signals transmitted through the coaxial cable, including those for two-way communications, have a frequency allocation as shown in Figure 6. The downstream transmission bandwidth of the coaxial cable system is from 70 to 250 MHz; there are fourteen signals in this cable for TV channels, two for facsimile channels, and some for the necessary control of the home terminal units.

Channel L, is used to convert and transmit off-the-air signal of Channel 10, from the antenna to the experiment center, for the flash information signal is super-posed on Channel 10 at the center.

The up-stream frequency band of the coaxial cable is less than 50 MHz; a TV channel is set up to relay original TV programme service.

The transmission line is set up to be able to transmit signals of good quality; for example, the following standards must be met:

cross-modulation .... less than - 46 dB

inter-modulation .... less than - 55 dB

carrier to noise ratio .... more than   42 dB

(for a video signal in the 4 MHz band)

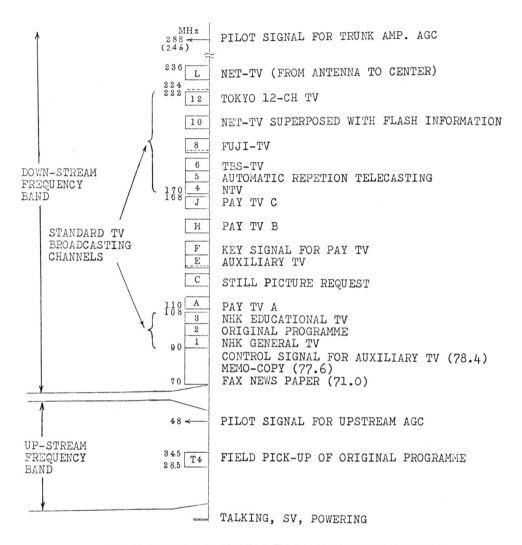

FIG. 6 FREQUENCY ALLOCATION IN COAXIAL CABLE SYSTEMS

4. Conclusion

The CCIS experiment is due to continue till September this year. Its final evaluation has yet to be made. A favourable response has been given to CCIS by the people in Tama New Town and great interest has been shown by the various parties concerned.

As a technical problem, some people are of the opinion that better picture quality should be developed in some services.

This problems resulted partly from the poor selectivity of the regular TV sets, since there are many input signal channels. Improvements have been made by properly adjusting regular TV sets. (The most frequently used services other than retransmission of off-the-air TV signals are the original TV programme service, the broadcast and response service, and the flash information service.)

Off-the-air TV programmes are mainly for recreation and entertainment, but it seems that a different attitude is being established in the monitoring households through CCIS services:

People have started thinking that CCIS programmes are not only "familiar TV programmes" but also "a new information system in which they can actively participate." They also feel that CCIS programmes would be much better if the content were enriched.

This kind of response can be easily surmised, judging from the high appraisal monitoring households have given to CCIS. In our investigation, we discovered that people feel each service is very useful and they want to continue receiving the services.

In terms of the needs of the people for various information, the information provided by CCIS is characterized by the following points: first, it is not only "the information we want to know", but also the information intrinsically connected to our daily lives and practically utilized as the occasion demands.

Second, the difference between the off-the-air TV programmes and the CCIS information is that our participation, both in mind and action, is one of the requirements for the latter.

Third, we must get used to the CCIS hardware such as terminal units. The expectation for software will be realized as actual needs only if it is deeply rooted in our daily lives. We have recognized through our experiment that the needs of the people for information stem from familiarity and not from coercion or unilateral service.

Kabelfernseh-Informationssystem mit interaktiven Diensten

Nach eineinhalbjährigem Studium der vom Kabelfernsehen als kommunalem Informationssystem angebotenen Möglichkeiten beschloß die Forschungsgruppe im Ministerium für Post und Fernmeldewesen, dieses System im tatsächlichen Einsatz zu erproben.

Dazu wurde ein Koaxialkabel-Informationssystem (CCIS) in Tama New Town, einem Vorort von Tokio, installiert und im Januar 1976 in Betrieb genommen. In diesem Pilotprojekt können eine Reihe neuer, auch interaktiver Dienste der Öffentlichkeit vorgeführt werden.

Die Teilnehmer stehen dem Probebetrieb, der noch bis September dieses Jahres andauert, grundsätzlich positiv gegenüber.

Der vorliegende Beitrag gibt eine Übersicht über die Dienste und den mechanischen Aufbau des Systems. Die Ergebnisse des Versuchsbetriebs werden nach dessen Abschluß im Detail veröffentlicht.

Hi-OVIS (Higashi Ikoma Optical Visual Information System) Development Project

Masahiro Kawahata
Tokyo, Japan

Abstract

The Hi-OVIS project is a government supported experimental trial for practical field use of an optical fiber communication system designed for the transmission of analog video signals and digital data signals between the terminal subscribers and the computer control center.
The trial consists of two phases:

In phase 1 the feasibility of the hardware and system design scheme was demonstrated by system prototype.

In phase 2 the optical fiber information transmission system is to be expanded to practical use. The system expansion is scheduled by step-by-step basis.

I. GENERAL REMARKS

The Hi-OVIS project is an experimental trial for practical field use of an optical fiber communication system designed for the transmission of analog video signals to subscribers and digital data signals between the terminal subscribers and the computer control center. The trial consists of two phases.

In Phase I, the feasibility of the hardware and system design scheme is to be demonstrated by system prototype. This prototype consists of the computer control center, optical fiber transmission system with video switches, and optical interfaces with the computer control center and the subscriber terminals. Phase I was completed November 1976 for technical feasibility demonstration of the in-building prototype in Tokyo, and is scheduled to be tested for field trial in the Higashi Ikoma area of a model city in the suburbs of Osaka, located about 300 miles west of Tokyo.

In Phase II, the optical fiber information transmission system is to be expanded to a practical use. The system expansion is scheduled by step-by-step basis:

Step 1   Expansion to 120 subscribers

Step 2 Expansion to 300 subscribers
Step 3 More than 300 in order to be economically justified

The 300-subscriber system design was completed by VISDA in September 1976, and detailed design including all equipments required for the 120-subscriber system (Step 1) was finished in March 1977. As far as Step 3 is concerned, we are exploring the possibility of expanding as profitable business from various standpoints such as market analysis, profit and loss analysis and organization.

Development Schedule is depicted in Figure 1.

II. HARDWARE SYSTEM DESCRIPTION

The system block diagram is shown in Figure 2.
As Figure 2 shows, the system can be divided into three subsystems; center, optical transmission line and home terminal system.

1) The center equipments are:

    i) TV retransmission equipment (VHF/UHF)
    ii) Local origination equipments
    iii) Equipments for request services such as AVCR, VCR, video cassette storage and retrieval system, etc.
    iv) Equipments for still picture and character information services such as character generator, still picture storage (micro-fiche) and voice information storage
    v) Control system for the equipments above and for handling subscriber's requests for various types of services; PFU 400 x 2
    vi) Computer system for statistical analysis and other business application; PFU 300 x 2

2) The optical transmission line consisting;

    i) Downstream video switcher and
    ii) Upstream video switcher
    iii) Opt-electro/electro-opt conversion device
    iv) Optical fiber cable

3) The home terminal system consists of;
    i) TV set
    ii) Keyboard
    iii) Camera
    iv) Microphone
      and in addition to the above
    v) Terminal controllers which controls the signal to and from transmission line and home terminals (TV set, keyboard, camera, and microphone)

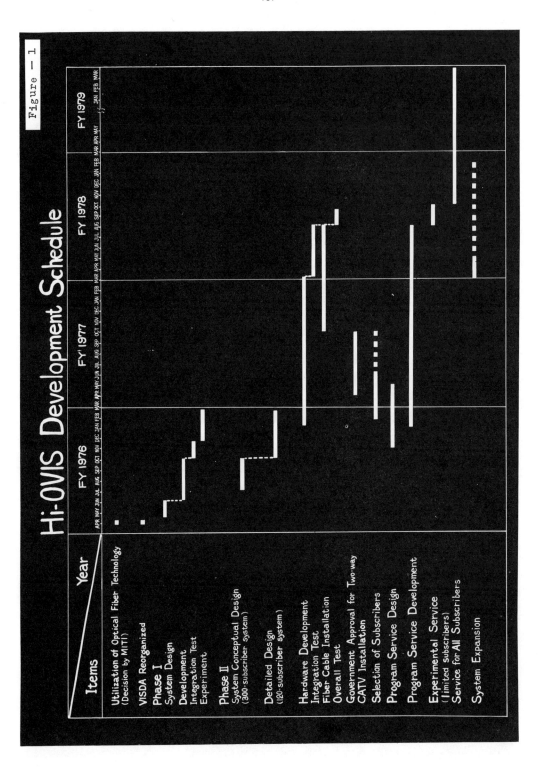

Hi-OVIS Development Schedule

Figure — 1

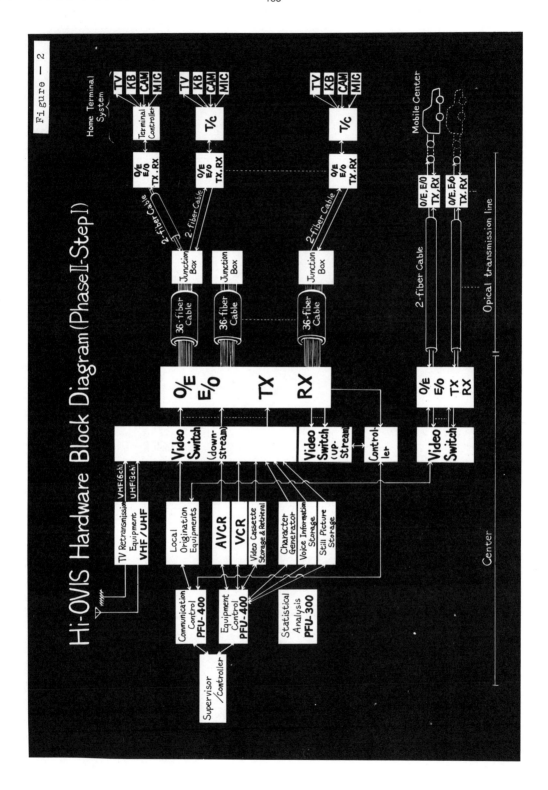

Figure — 2

Hi-OVIS Hardware Block Diagram (Phase II-Step 1)

III. BASIC CONCEPT FOR PROGRAM SERVICE DEVELOPMENT

1) New community development

Because of increasing complexity of the social structure and diversity of appreciation of the value, the traditional local community has been destroyed, however, there still exists strong needs for the development of the new community. The Hi-OVIS program service is to be designed to meet this requirement.

2) Service for individual oriented information selection

Complexity of the society, progress of technology and cultural changes have caused individual to get his own subjective information through the system with the capability of retrieving audio-visual information. The Hi-OVIS is capable of satisfying these needs.

3) Needs for life long education

Program service includes various types of educational program in order to catch up and well adapt to ever-changing social environment.

4) Approach for social welfare

Program service development is especially based on old people's welfare problem such as medical care and social activities.

Along with the concept above, channel allocation for various services are planned as follows:

  i) Retransmission of on air TV services - 9 channels

NHK
NHK (Education Channel)
Kansai TV   VHF
Mainichi TV
Asahi TV
Yomiuri TV

Sun TV
Kinki TV   UHF
Nara TV

 ii) Local origination - 1 channel

Two-way interactive service using keyboard, TV camera (black & white) and microphone located at subscriber's site are the most outstanding feature of this service.

In addition, two mobil centers with two-way capability are available for relaying special events or information from local school, municipal offices, shopping districts and so on. Examples of the local origination service are:

a) Wide show including consultation and subscriber participation services, local news, weather forecast, shopping information, traffic information, local activity notice, subscriber's produced program, subscriber's performed program, etc.

b) Educational service which includes;

Two-way TV study consultation
Two-way TV summer school, etc.

c) Local activity participation such as;

Athletic meeting
School excursion
Local election
TV discussion, etc.

iii) Request Services (Video Cassette Retrieval) - 7 channels

A wide range of programs such as education, culture, hobby, movie, drama and sports relays are able to be requested through keyboard. Center hardware equipments for the service are video cassette storage and retrieval system (capacity 60 cassettes), six VCR's and one automatic video cassette changer. 7 channels are allocated for this service:

a) Three arbitrary request service channels (using video cassette storage and retrieval system)

b) Four selection channels (using six VCR's and one automatic video cassette changer)

iv) Still picture and character information service - 7 channels

The character information is served by character generator system controlled by computer. Advantages of this service are:

a) Information is easily updated.
b) Response time is comparatively short.

The character includes KATAKANA, HIRAGANA, KANJI, alpha-numeric with seven kinds of coloring capability.

The character information covers;

a) Local news
b) Highway traffic information

    c) Weather forecast

    d) Railroad/airline time table

    e) Hi-OVIS channel guide

The still picture information is to be stored in the form of micro-fiche in order to serve educational, cultural or welfare information according to the request from home terminal (keyboard).

Six channels are for character information service and the remaining one is dedicated for still picture service.

IV. PROJECT BUDGET & COOPERATING COMPANIES

Approximately US$9.4 million is going to be invested for Phase I and Phase II step 1. Additional investment (several million dollars) is expected for future expansion.

Three companies are responsible for supplying hardware equipments. Matsushita Electric is responsible for center audio-visual equipments as well as home terminal system. Sumitomo Electric is responsible for analog video transmission lines including optical fiber cable, O/E or E/O device and video switcher. Fujitsu is responsible for center computer system (both hardware and software) and O/E or E/O device for digital link between home terminal and center computer.

As for program service development, Fuji Telecasting Co. awarded contract for the development.

In addition, Dentsu Advertising Co. is engaged in Hi-OVIS promotional activities (dedicated to local Higashi Ikoma community) for VISDA.

V. PHASE-IN TO THE COMMUNITY AS A WELL-ACCEPTED SOCIAL SERVICE SYSTEM

One of the ultimate objectives of Hi-OVIS is to build a social information service system which is to be well-accepted by modern communities, with a high density of use and transfer of information. Exploitation of optical fiber cable for this information service system is also very important in view of the many advantages which can be provided by optical fiber cable transmission systems. Some of the major goals of this experimental trial include the following:

    1) Low cost high capacity transmission lines (big potential of fiber optic technology)

    2) Low cost subscriber terminal

    3) Easy-to-expand system architecture (sub-center concept)

    4) Community involvement for system operation

    5) Justification of Higashi Ikoma system as prototype for a future social service system in Japan

    6) Prospect of Higashi Ikoma system in future fiber cable-based wired communities.

ACKNOWLEDGEMENTS

The author would like to thank Messrs. T. Sakai, T. Arai and Y. Honda of MITI for helpful advice and discussions for the project and the results presented represent the combined effort of Mr. S. Iijima and other many individuals at VISDA to whom the author is deeply grateful.

REFERENCE

M. Kawahata, "Development of Optical Information Transmission System - Field Experiment in Higashi Ikoma", presented before the Third European Electro-optics Conference Geneva, Switzerland, October 5, 1976.

Das HI-OVIS-Entwicklungsprojekt (Optisches Fernseh-Informationssystem in Higashi-Ikoma)

Das HI-OVIS-Projekt ist ein staatlich gefördertes Versuchsvorhaben zur praktischen Einsatzerprobung eines Zweiweg-Telekommunikationssystems mit optischer Übertragung auf Glasfaserkabeln, das für die Übermittlung analoger Video- und digitaler Datensignale zwischen den Teilnehmern und einer rechnergesteuerten Zentrale konzipiert wurde.

Der Versuch verläuft in zwei Phasen:

In Phase 1 wird die Realisierbarkeit der Hardware und des Systemkonzepts anhand eines Prototyps dargestellt. Dieser Schritt wurde im November 1976 abgeschlossen.

In der jetzt begonnenen Phase 2 wird das Versuchssystem auf den praktischen Einsatz ausgeweitet, wobei max. 300 Teilnehmer angeschlossen werden sollen. Man hofft, wichtige Erfahrungen aus dieser "Modellstadt einer zukünftigen informierten Gesellschaft" zu gewinnen.

Die Arbeit gibt eine Übersicht über die Planung dieses Systems, die darin vorgesehenen Dienste sowie die notwendigen Einrichtungen beim Teilnehmer und in der Zentrale. Jeder Teilnehmer ist mit je einer Glasfaser für die Vorwärts- und einer für die Rückwärtsrichtung an eine von mehreren Unterzentralen angeschlossen, in denen eine Breitbandvermittlung die Verbindung zwischen dem Teilnehmer und den von der Zentrale verteilten Programmen herstellt. Insgesamt sollen etwa 500 km Glasfasern in diesem System installiert werden.

Video Information System for IR and CAI Using Telephone Pair Cables and Home TV Sets

Hirohito Nakajima
Tokyo, Japan

ABSTRACT

An experimental video information system for information retrieval (IR) and CAI for internal NTT utilization was cutover on 10
January 1977. It is projected as a first step toward developing
versatile communication media with individualized audio and video
communication.

This trial system is composed of an information file control
center, about 100 terminals and individualized transmission lines.
Several kinds of service are contained in this trial system, so as
to allow evaluating system utility and functions.

The features and outline of the experimental system are described in this paper.

1. INTRODUCTION

In video communications, such system as CATV, video conference
and video telephone have been developed under NTT auspices.
Among them, two-way broad band multi-function CATV systems were
developed after 1970. The latest system, now applied to Tama Coaxial
Cable Information System, provides 27 downstream TV channels, 5 upstream TV channels and several channels for sound program, data and
facsimile through a coaxial pair utilizing VHF band carriers.

The CATV system, based on retransmitting a conventional television program, allows instigating many functions with reverse
channel, such as locally originated programs, still picture information retrieval upon request and so forth. Therfore, this system

is being used in various communities as a means of providing local information.

Considering such versatile service in the future, an interactive video information system will be urgently required. It has been recently considered in NTT that center-to-end type interactive systems are important as a future communication media and such an experimental system was developed for the first step, whose internal NTT utilization service was initiated 10 January 1977.

2. GENERAL FEATURES OF THE CENTER-TO-END TYPE INTERACTIVE SYSTEM USING EXISTING TELEPHONE CABLES

It is considered that the most adaptable and realizable center-to-end type interactive system is to be composed of an individualized network using existing pair cables with video band repeaters. Such a system has following features.

(1) System utility does not depend so much on the number of subscribers as in the case of end-to-end type system.

(2) Each subscriber can independently select abundant information in the form of video signal by taking advantage of individually distributed network.

(3) Each subscriber can control program sequences by means of interactive functions between the center and the subscriber.

Such services as information retrieval (IR) and computer assisted instruction (CAI) are able to be furnished naturally with audio and video including still and moving pictures.

3. EXPERIMENT OBJECTIVES

Experimental system objectives are as follows:

(1) Investigation on system functions and total performance

(2) Hardware verification

(3) Study on software constitution

(4) Study on service materials and their hardware interface

(5)   System construction and maintenance

(6)   System utility and cost

(7)   Determining technical problems for the future

4.   EXPERIMENTAL SYSTEM OUTLINE

A basic configuration of the experimental video information
system is, as illustrated in Fig.1, composed of an information file
control center, about 100 terminals and individualized transmission
lines using ordinary telephone pair cables and broad band pair cables
with video band repeaters.

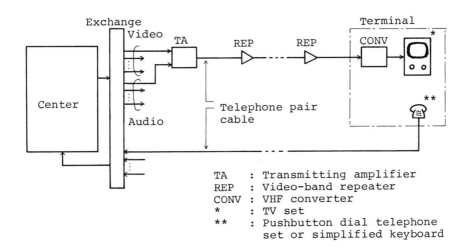

TA   : Transmitting amplifier
REP : Video-band repeater
CONV : VHF converter
*   : TV set
**   : Pushbutton dial telephone
       set or simplified keyboard

Fig. 1 Experimental Video Information System
        Basic Configuration

In Fig.1, a requesting signal from a pushbutton dial telephone
set or simplified keyboard at a terminal end is received by a pushbutton signal receiver through the switching equipment in the center.
After translating the requesting signal, an information file, that is,
picture and sound data corresponding to that requested is selected and
sent to a terminal under program control.   The audio and video baseband signal transmitted through several repeaters is converted to VHF

TV signal by a converter at the terminal end. The picture is supplied
to a vacant TV channel of a conventional TV receiver.

Fig.2 shows the center configuration in more detail. One frame
of a still picture selected from a picture file is transferred to a
frame memory. If a frame is to be a composit of several picture
files, the composit is set up on the frame memory. After transferring
requested data to a frame memory, the picture file is released to wait
for another incoming call.

In this system, a series of one-frame still picture is repeatedly
transmitted to a terminal through the switching equipment and combined with the audio signal at a transmitting amplifier.

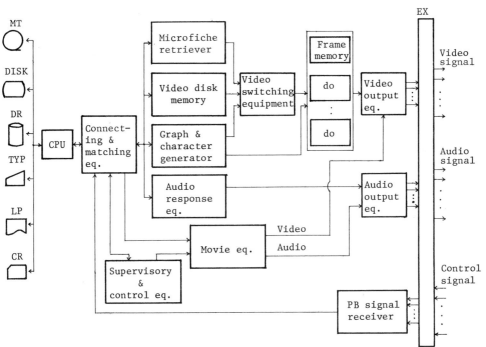

Fig.2 Experimental Center Configuration

5. CONFIGURATION TECHNIQUE

(1) Experimental network scheme

The experimental video center is installed at a telephone
office. About 100 terminal sets are connected to each telephone

office or other NTT office building about three kilometers from
the center. Each terminal set is individually connected to the
center by a direct transmission line, these 100 lines are concentrated to 20 lines inside the center.

(2) Transmission line

The downstream line for the audio and video transmission
handles baseband transmission of a 4 MHz NTSC television signal
combined with audio signal modulated to higher frequency allocation over telephone pair cable or twin type broad band
communication cable with video band repeaters.

The repeater span is about 0.5 km for telephone cable and
about 2.0 km for broad band cable, respectively.

The upstream line, to send a request signal from a terminal
to the center, is an ordinary telephone cable itself with no
repeater.

The signal is ordinary pushbutton signal itself.

(3) Terminals

A terminal consists of a television receiver, additional
converter and simplified keyboard or pushbutton dial telephone.

The converter changes received baseband signal to VHF band,
so it can be received in a vacant VHF TV broadcast channel.

There is also a means to switch baseband signal on and to
eliminate any TV signal leaking from some other TV broadcast
channels, by detecting a horizontal synchronization signal of the
baseband signal.

The simplified keyboard or pushbutton dial telephone is used
for requesting information and controlling program sequence. The
keyboard is designed for easy use with four extra functional
pushbuttons, such as backward control, returning, repeating and
cancellation.

(4)   Experimental center

The center consists of several kinds of picture file equipments, video frame memory devices, control computer, etc.

Microfiche retriever (MR), graph and character generator
(GCG) and video disk memory (VD) are provided for random accessible still pictures. An outline of each information file equipment is shown in Table 1. Their capacities are as follows:

```
MR  :  6,000 color picture
GCG : 10,000 frames
VD  :    900 color pictures
```

A 16 m/m television projector and video tape recorder are
used for the broadcast type moving picture service.

The audio response equipment has a random accessible file
as a sound source. Replies are controlled by the editing method,
whereby a sentence is comprised from about 1 second long words
stored on a magnetic drum, formed into pulse code modulation.

Another type of audio file is provided for announcement of
the system operating status to calls, such as "closed",
"troubled" and so on.

Table 1   Still Picture File Outline

Equipment	Capacity/Eq.	Average access time	Notes
Microfiche retriever	6,000 frames	3.8 second	
Video disk memory	900 frames	1.0 second	Including animation
Graph and character generator	10,000 frames depend on the external memory data area	Character 30 ms/100 ch. Graph 10 ms/10 dots	Size of character standard size 32x32 dots small size 20x20 dots Kinds of character 2,300 including chinese ideograph

Video frame memory units, using a 10 cm diameter magnetic disk, are equipped for each still picture trunk circuit.

Furthermore, crossbar switching equipment is provided for this experimental system to allow a voice and a video switching frame to be submitted to it separately.

These audio and video files and other associated equipment are controlled by a stored program using a small size computer system. The system program configuration is shown in Table 2.

Table 2.   System Program Structure

Executive control program	—	Input/Output control program

- Communication control program
- Administration program
- Fault processing program
- Movie control program

- Job management program
- Journal logging program
- File utility program
- Language compiler
- Graph & character edit program

6.   SERVICE SPECIFICATIONS

Two types of service are available, still picture service and movie service.

The still picture service is advanced interactively. There are nine different input requesting functions. These requesting functions and operations are explained in Table 3. Input numerical data are indicated on the lower left corner of the TV screen, and can be monitored. Output picture and sound are presented automatically, after selection in response to the input signal.

Table 3. Request Functions & Operations

Functions	Contents	Operations	
		PB dial telephone	Keyboard
Service start	Connect to the center and start a service	Off-hook & dialing	Power-on & dialing
Service end	Release the line and end a service	On-hook	Power-off
Input end	End the input data to the center	#️⃣	#️⃣
Next	Go to next frame	#️⃣	#️⃣
Repeat	Repeat the frame	7️⃣ *️⃣ #️⃣	⇧
Back	Back to the last frame	8️⃣ *️⃣ #️⃣	↩
Return	Return to the backward beginning frame	9️⃣ *️⃣ #️⃣	↵
Cancel	Cancel input data	0️⃣ *️⃣ #️⃣	✕
Stop	Stop service and request other service	*️⃣ *️⃣ #️⃣	*️⃣ *️⃣ #️⃣

The movie service, which is requested by a different code from the still picture code, is now provided for the time scheduled "broadcast" only. All the time, except on movie service, their program guide with still picture is available.

The service materials used in this system are produced only for NTT inside use experimentally. Their contents, devided into several categories, are shown in Table 4. Some other services such as primary computer technique and practical English conversation are to be added soon.

Table 4. Experimental Service Material Outline

Category	Service materials	Notes
Studying by CAI method	English grammer for beginners 1~3	
	Drill in mathematics 1~10	
Guidance by IR method	Telephone sales guide	
	Guide to public inns and lodges	
Quiz and learn	Table manners in western and Japanese style	
	History in region	
	Quizzes	to be added
Game	How to play "go" 1~14	by CAI method
	Dolphin shooting etc.	
Movie	Business knowledge, culture, etc.	

7. CONCLUSION

The services offered by this experimental system are not sufficient because its service materials were produced by NTT engineers themselves. This first experimental system still has much room for improvements. Some of these improvements are scheduled for the next system. Furthermore, combined still and moving pictures will be studied.

The utility of this type of system depends mainly on both kinds and contents of services. It is important to take consecutive efforts to develop hardware and software through system utility verification on an actual system.

Fernseh-Informationssystem für Informationswiedergewinnung und rechnergestützten Unterricht unter Verwendung von zweiadrigen Fernsprechleitungen und Heimfernsehgeräten

Die Versuchsanlage eines Fernsehinformationssystems mit IR (Informationswiedergewinnung) und CAI (rechnergestützter Unterricht) für den betriebsinternen Einsatz wurde am 10. Januar 1977 bei NTT in Betrieb genommen. Die Anlage ist als ein erster Schritt bei der Entwicklung von universell einsetzbaren Kommunikationssystemen mit individuellem Bild- und Tonverkehr gedacht.

Das Versuchssystem besteht aus einem Datensteuerzentrum, ca. 100 Endgeräten und individuellen Übertragungsleitungen. In diesem Versuchssystem werden verschiedene Arten von Diensten zusammengefaßt, die eine Auswertung der Systemdienste und Funktionen ermöglichen.

Merkmale und Aufbau der Versuchsanlage werden im vorliegenden Beitrag beschrieben.

Developments in Dial-Access T.V. Systems and Fibre Optic T.V. Transmission

K. C. Quinton
Kingston-upon-Thames, Great Britain

GENERAL

This paper provides a brief account of developments in the United Kingdom during the last 8 years of t.v. distribution systems which are based on the principle of having a primary cable network feeding exchanges and individual lines between the exchanges and domestic outlets.   The exchange switching and required radius of unrepeatered coverage from the exchange becomes practicable if transmission at high frequency is adopted. The program capacity of the system is independent of the secondary network, thus any future requirements can be accommodated by augmenting the primary network and adding further program selectors. Provision is made for 2-way t.v. transmission on the outlet lines.

The transmission of one composite t.v. channel in each glass fibre, applying and receiving at the terminations h.f. signals identical to those in use in exchanges, has been proved on two 20 dB fibre links commissioned in March 1976. The author is confident that when glass fibre cables become cheaper than copper cables they will replace the latter on both primary and secondary links of the dial - access system. Fibres for the secondary links can be of a comparatively low grade.

ELECTRO-MECHANICAL EXCHANGES

In 1966, my company was considering cable t.v. in the rural environment, and we concluded that the best way to serve a dwelling which was remote from a village was to make connection by a screened quad cable and to provide program selection by remote control of a switch at the sending end. By using a vision carrier frequency of 8 to 9 MHz and cable with 0.7 mm diameter conductors, we could, on the subscriber cable, adopt an amplifier spacing of at least $1\frac{1}{2}$ km and we could use the second pair in the cable for the remote control.   Following this principle, we set about designing a t.v. equivalent of a small telephone exchange operating at high frequency.   Our exchange was designed to feed 336 outlets from a choice of 36 input programs, and we chose reed relays for the switching operation.

Figure 1 illustrates the switch which provides 2 independent outlets from 36 inputs. The transmission lines to and from the relays are strips sandwiched between earthed copper areas, and the reeds are operated by a ferrite permanent magnet on an arm, which is turned by a mechanism broadly similar to that in a Strowger telephone selector. There are small ferrous shields between the reeds, and it is interesting to note that access to any particular channel can be denied by placing a small ferrous clip over the appropriate reed so that it will not be closed when the magnet is above it. Pay t.v. on a per channel basis can be arranged using this principle. The selector is operated by a standard telephone dial at 10 pulses per second, and the unbalanced t.v. and sound signals are converted to the balanced form for connection to the outgoing pair cable.

Figure 1

Several exchanges made to this design are currently in operation, and in 2 locations a pair of exchanges is used to provide increased facilities. In the S.A.B.C. studio and office complex in Johannesburg access to 72 channels is provided, and in the Nova Park Conference Centre in Zurich 556 outlets are provided with access to simultaneous conferences, to off-air programs and to a bank of 54 video cassette machines. Figure 2 shows the Zurich exchanges.

Figure 2

Reed relays are still being introduced to telephone exchanges, and we have found them to be extremely reliable; indeed, so far we have had no operational failure of a reed or of a mechanical selector on many exchanges which have been in operation for some years.

SOLID STATE SWITCHING

There are many areas where 36 channel capability is not immediately required, and it can be an economic proposition to install a reduced capacity and to add to it when required. Our calculations show that for 24 channels or less an exchange based on a solid state matrix switch is less expensive than one based on reed relays. Figure 3 shows the copper track side of a 48 crosspoint switch which provides 4 outlets with access to 12 channels. Fifty of these switches fit into an exchange which is small enough for installation as a roadside cabinet, see Figure 4. Preparations are in an advanced stage for the installation of 5 solid state exchanges in the Netherlands.

In both types of exchange which have been described signal amplification precedes the bus-bars, and the switches operate at line sending level. The semiconductor switch, which has shift registers on the same card, will accept pulses at 200 per second, which are delivered from a touch-button pad without signal storage being required.

Figure 3

Figure 4

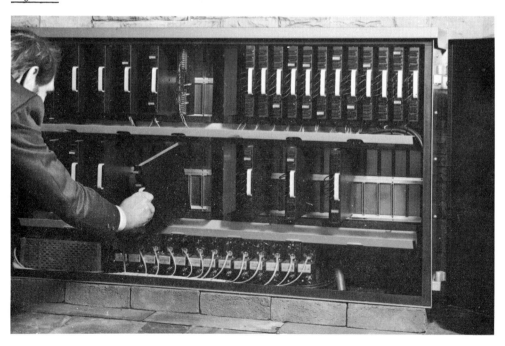

The outgoing cables are to a special design, registered as Qwist, which gives very good crosstalk protection at high frequencies. Types available have capacities from 1 to 12 outlets within a common sheath, each outlet being a 4-wire connection, and Figure 5 shows the cross-section of a 6-outlet cable. The maximum range using 0.45 mm vision conductors is 400 metres, which suits urban installation.

Figure 5

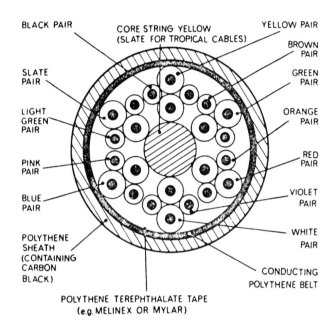

There are several ways in which the 4-wire connection can be used. One is to put the selected vision signal on the larger pair together with power for the domestic frequency converter to v.h.f. or u.h.f., and the smaller pair would then carry dialling pulses and f.m. stereo channels which are distributed using f.d.m. at h.f. since they are not selected at the exchange. An alternative is to transfer the dialling pulses to the vision pair, using a low carrier frequency if converter powering by line is necessary, releasing the smaller pair for the normal telephone function but adding to it the f.d.m. radio channels at the exchange. The exchange would be used as a telephone cabinet so far as the small pair is concerned.

FACILITIES AVAILABLE

If required, the t.v. pair connected to each outlet will carry additionally an upstream t.v. signal on an h.f. carrier frequency which is twice that of the downstream signal. The upstream signal is converted to an exchange input by a frequency changer located within the exchange, and this signal is also available for connection to the primary network of inter-exchange lines. Upstream transmission using this principle occurs frequently on the dial - access system installed in the Case Western Reserve University in Cleveland, Ohio. The principle is capable of extension to videophone use should this be a requirement.

Once an exchange principle is adopted, denial of access to particular programs can be provided by switches in the exchange. These can be set manually for special audiences, such as doctors or police, or for per channel Pay t.v., or they can be controlled from a centre for per program Pay t.v. which could be computer controlled and billed at an adjustable rate if required.

The above descriptions are very brief, and further information can be obtained from published articles, references :

1. Gabriel, R. P. :   Two-way experience with Dial-a-Program at Dennis Port.   21st Annual N.C.T.A. Convention 1972. Official Transcript Technical Volume, pp. 375-383.

2. Gargini, E. J. :   The total communication concept for the future.   Royal Television Society Journal, 1973, March/April, pp. 182-193.

3. Gargini, E. J. :   Solid state switched cable television system.   Proceedings of Eurocon '77, Volume 1, pp. 2.11.81-86.

4. Pacey, J. F. :   Applications of the Dial-a-Program system - with particular reference to the Nova-Park Hotel and Business Centre, Zurich.   Cable Television Engineering, 1973, Oct., pp. 61-74.

FIBRE OPTICS IN THE PRIMARY NETWORK

It is my belief that the cable conductors in a dial-access system can easily be replaced by optical fibres, whereas the application of fibres to tree and branch types of network is unlikely to occur in the near future, if at all. This is because of the difficulties in operating an l.e.d. with several input channels in f.d.m., partly because of non-linearity and because of the very small optical power per channel which can be launched. We have shown that an optical fibre link is a very practical proposition if the composite t.v. channel is applied and received using h.f. carriers which can be handled in an exchange without crosstalk and isolation difficulties.

To prove the operation of a fibre optic link with h.f. input and output, we installed a 1½ km 2-fibre cable in England in 1975, and this has been in service since March 1976. Figure 6 shows the route of the cable, which was drawn through 3 inch salt glaze ducting which already contained up to 14 cables previously installed. The two t.v. signals, one from each fibre, are subsequently distributed to over 34,000 homes.

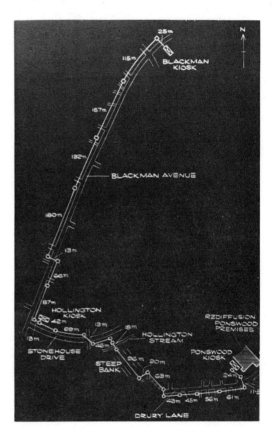

Figure 6

☐   Manhole
⊠   Kiosk
☰   Footpath
      Metalled Road

Figure 7 illustrates the cable, which was made by B.I.C.C., and the fibres are Corning step index type having an attenuation of 14 dB/km at a wavelength of 900 nanometres. The cable size is 7 mm x 4 mm, and the fibres are illustrated passing through a No. 5 sewing needle. The fibres rest loosely in a cavity such that they are on a neutral axis when the cable is flexed. Two steel strength members limit the cable extension whilst drawing to well below the 1% extension which would break the fibres. The sending device on each fibre is a Plessey Burrus diode giving a radiance of 50 watts / steradian / cm^2, and it delivers about one-eighth of a milliwatt into the 85 micron core of the fibre. As there are no repeaters, the received power is 20 dB lower, i.e. 1.2 microwatts.

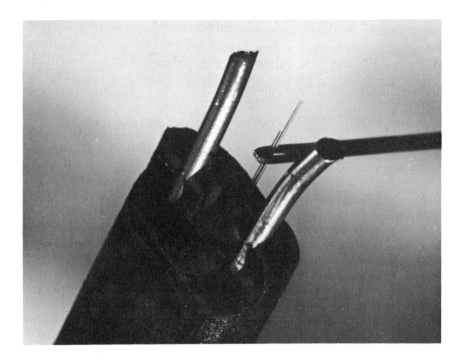

Figure 7

It should be noted that the sending diode and the PIN receiving diode are devices for converting electric current to light and vice versa. It follows that the effect of the insertion of a 20 dB fibre between these devices is roughly equivalent to that of the insertion of a 40 dB piece of conventional cable.

With this link we achieve a "just perceptible" noise contribution to the picture, but cables now available have a much lower fibre attenuation per unit length and the cascade operation of many $1\frac{1}{2}$ km fibre sections is now a practicable proposition. Joints in the cable are made by impressing a piece of fibre into soft copper and then butt jointing prepared ends, using the indentation in the copper for alignment, and clamping the fibres in position.

FIBRE OPTICS IN THE SECONDARY NETWORK

There are no problems in using optical fibres between exchanges, but the ideal fibre for subscriber lines should have different characteristics to the type used for long distance telecommunications purposes.

The optical power which can be launched into a fibre is proportional to the area of the fibre core and to the square of the numerical aperture. Thus, by using a fibre with a core diameter of 280 microns and with a numerical aperture of .38 (compared with .16 for the Corning fibre described previously), the light input can be increased by 50 times. Thus, we envisage subscriber lines from an exchange based on larger fibres with increased aperture which permits the use of less expensive l.e.d.'s, since the cost increases as the diameter of the light emitting surface decreases. We are currently developing a cable which will carry one or two comparatively large fibres for the secondary network. An increased numerical aperture implies more optical dispersion, hence a lower product of length times bandwidth, but this is acceptable for the lengths envisaged. Ideally we should use all-plastic fibres as, being more robust, they are better suited to domestic installation. So far, the attenuation of these fibres is too high for them to be considered.

In all instances we anticipate extraction of analogue h.f. signals from the fibre circuit and simple frequency conversion to a recognized broadcast t.v. channel.

Entwicklungen auf dem Gebiet der HF-Kabelfernsehsysteme mit Wählzugriff und der Video-Übertragung über Glasfaserkabel

Die Arbeit gibt eine kurze Darstellung der Entwicklung von Fernseh-Verteilsystemen in Großbritannien, bei denen die Fernsehprogramme über ein Primär-Kabelnetz den Verteilzentralen mit ihren Vermittlungen zugeführt und von dort über individuelle Leitungen an die Teilnehmer verteilt werden. Die Verteilung sowie der erforderliche Radius des von der Verteilstelle aus ohne Zwischenverstärkung zu versorgenden Gebiets können durch HF-Übertragung realisiert werden. Die Kapazität des Systems ist unabhängig vom Sekundärnetz, wodurch spätere, zusätzlich auftretende Anforderungen durch Erweiterung des Primärnetzes und Einbau weiterer Programmselektoren berücksichtigt werden können. An den Teilnehmeranschlüssen sind Möglichkeiten für die Zweiweg-Fernsehübertragung vorgesehen.

Die Übertragung eines Fernsehkanals in jeder Glasfaser, wobei die gleichen HF-Signale wie in den Verteilstellen von den Endanschlüssen aus gesendet bzw. empfangen werden, wurde an zwei im März 1976 in Betrieb genommenen 20 dB Glasfaserkabelstrecken erprobt. Der Verfasser ist davon überzeugt, daß Glasfaserkabel, liegen deren Kosten erst unter denen der Kupferkabel, sowohl in Primär- als auch Sekundärverbindungen dieses Systems eingesetzt werden. Dabei können die Glasfaserkabel für das Sekundärnetz von relativ geringer Qualität sein.

The Digital Return Channel
on the Rennes Experimental Cable Television Network

C. Sechet
Rennes, France

Abstract

The TIGRE system, which was designed for cable TV networks
with a return channel, gathers together on such networks logical
states describing terminals such as TV programme selectors of subscribers, and processes them by means of a central mini-computer,
thus providing accountancy services on the use of certain special
programmes and remote water-meter reading. Numerous other uses may
be envisaged, such as the possibility of dialogue between TV viewers
and programme organisers.

I - INTRODUCTION

The C.C.E.T.T. is experimenting in Rennes with a prototype
cable TV network whose significance lies in its ability to transmit
15 TV and 10 radio bands and also in the addition of a return channel
to this normal broadcasting channel.

Frequencies transmitted on the cable are divided into two
bands : one is the outgoing channel which operates from 50 to 300 MHz
and the other the return channel which operates between 5 and 30 MHz.
The TIGRE digital system makes use of this latter link. It provides a
digital return channel.

II - FUNCTIONS OF THE DIGITAL RETURN CHANNEL SYSTEM

The TIGRE system uses two-way transmission of digital signals
along the same coaxial as TV signals for most links, and which joins
up a mini-computer at a central station with a collection of concentra-

tors (the Star Network Data Collection Terminals*) situated near the furthest points of the network, being placed for example at the foot of blocks of flats containing some hundred subscribers.

*Translator's Note : For convenience the French term TIGRE (Terminaux Informatiques pour Groupe Raccordé en Etoile - Star Network Data Collection Terminals) has been retained throughout this text.

A TIGRE concentrator comprises firstly a section for exchanges with the central station, providing the data feeding service, as well as a section to read the states of the different subscribers' terminals, known as data pick-up service.

II.1 - The DATA TRANSMISSION OF FEEDER SERVICE

The sub-group, which includes a central station type modem opposite a concentrator type modem, and the logic elements controlling the communication procedure make up a channel which is transparent for useful data travelling in both directions in 4-byte packets.

We there fore have at our disposal a band with a 70 kilobit useable capacity in either direction, to be divided among its users according to circumstances.

II.2 - SERVICES WHICH THE PRESENT VERSION CAN PROVIDE

The present TIGRE system only makes use of the return transmission facility and the pick-ups are of a simplified design, capable of recognising only four states in the subscriber's terminal.

These are :

1st type : Link-up to a subscriber's programme selector providing information as to the state of certain switches. It can be used to monitor the actual use of 3 special TV programme channels and thus the charge to be invoiced, or to set up an elementary two-way communication.

The calculator monitors which of the 4 following states applies to the subscriber :
Code 00 - TV switched off or tuned to ordinary programmes
Code 01 - 1st special programme channel

Code 10 - 2nd special programme channel
Code 11 - 3rd special programme channel

This can correspond to the answers yes, no or may be. If the actual use of the special programmes is to be evaluated, then the selectors are scanned every minute as to their state.

2nd type : Link-up for remote reading of liquid flow meters.

Meters suitable for this facility already exist, and the experiment has been carried out using water meters. The calculator reads the open-shut state of a gate activated by the meter turning. To work out the consumption it adds up these activated states.

3rd type : Link-up to alarms (burglar, fire alarms, etc...)

III - GENERAL STRUCTURE

Fig. 1 shows the system as a whole.

In the central office there is a mini-computer - a T 1600 - joined to the coaxial network by a modem of which the outgoing channel (68.75 MHz) and the return channel (7,64 MHz) are multiplexed with the other radio and TV frequency modulations.

This mini-computer's task is to 'question' the TIGRE concentrators and to process appropriately the information coming in from them.

The cable TV network has a tree-network development. Near the extremities of the base network are to be found the concentrators in the form of cabinets measuring 45cm x 65cm x 20cm, limited by the address system to 500, which communicate on the one hand with the central station and on the other gather data from the 128 or 256 lines corresponding to that number of terminals in the form of TV programme selectors, meters or alarm systems. Data thus do not travel directly from the terminals to the central station, but are relayed by the concentrator.

In this way the number of sources using the return channel at any one time and capable of distrubing it are reduced from several thousand to 500. Moreover they can be situated in spots with easy access which greatly simplifies maintenance and, if necessary, repairs.

Further, Fig. 1 shows clearly that if the base network is
single, the distribution network is doubled. In the apartment blocks
the TV cabling is of the classic tree type, but the digital data travel
on a second star network from the concentrator. This option has the
advantage of allowing the coding element in the terminal to be as
uncomplicated as possible, and in certain cases of allowing TV programme distribution to be cut off, but these advantages are countered
obviously by a higher cable cost. It is particularly suitable then for
new buildings where it can be installed at the same time as the telephone and electricity cables.

IV - THE CONCENTRATOR

IV.1 - The lines and the subscriber pick-ups

Fig. 2 shows the schema of the link-up of a subscriber to the
concentrator. It is backed up by a two-wire telephone line constituting
a current loop. This loop is fed from a subscriber card in the TIGRE
by a low-frequency alternating current (about 600 Hz) at a voltage
between-4v and 4v. It is closed at the subscriber's end by contacts in
series with diodes. This very elementary method allows two independent
binary states to be coded, one contact for positive and the other for
negative current alternations.

On the subscriber card in the concentrator are the data pick-
ups, made up of transistors and translating the binary states into a
form allowing them to be operated on by the logic of the system.

A TIGRE concentrator gather two groups of 128 binary states
which may or may not be present by virtue of a modular principle, and
an address sent by the calculator refers to 16 at a time so that 16
'questions' put to the concentrator are enough to establish the state
of all 256 lines. No link is possible which would give the subscriber
terminal the possibility of sending a complex information message, and
it is only suited to the remote meter-reading function initially programmed.

Where the subscriber terminal is a TV programme selector, it should be noted that the digital coding signals are remultiplexed along the coaxial from the last junction so as to keep down the number of wires being led into the apartment.

IV.2 - The H.F. Link with the Calculator

The 500 concentrators transmit on the same channel (7.64 MHz) and pick up data at 68.75 MHz.

Temporal multiplexing is thus necessary, determined by a scanning cycle sent out from the calculator according to a question-and-reply formula for each address successively.

The messages are made up of 55 bits - which is short - which can be broken down for question and replies into 16 bits for the address of the concentrator and the group within it, followed by 32 data bits which in the return direction indicate the binary states of the 16 lines examined, but which in the outgoing direction are at the moment empty, and lastly a redundancy code for the purpose of error detection control.

The system used is 4-phase shift keying modulation in accordance with the conversion where by a '1' is represented by a variation of $+\pi/2$ and a '0' by $-\pi/2$. In this way the clock is transmitted during the jump instants and the binary symbol by their sign. It operates at a speed of 500 kilobauds.

But given the inevitable dead times and the structure of the messages, the most useful speed is around 70 kilobits per second, and even this is too much for present needs.

IV.3 - Internal reorganisation of a TIGRE (see Fig. 3)

A concentrator is essentially made up of the elements allowing it to communicate with the calculator, the modem, the serial-parallel, parallel-serial conversion register, sequence recognition and control, and those linking it to the subscriber, data pick-ups and multiplexors.

Every message picked up from the coaxial network is loaded into the
serial-parallel conversion register and the address and cyclic code are
examined.

From among the questions one is recognised to be intended for
the concentrator. If no error is revealed by the cyclic code, then the
address of the group present in the register directly takes over the
multiplexor and the 32 bits wanted are introduced to make up the reply,
which is then transmitted by means of the following question's clock as
soon as it comes through, whether or not that question is intedned for
the concentrator. The nth reply is communicated to the calculator after
the $(n + 1)^{th}$ question.

It may be noted also here that there is no local clock but that
on the contrary everything is synchronised by the central station clock,
which considerably simplifies the TIGRE and the data reception elements.

V - THE CENTRAL STATION

V.1 - The hardware

The TIGREs effect a first concentration of the digital data
to be collected. The system operating on the data from the digital
channels in the network should firstly effect a second concentration
corresponding to the reception of data at the central station and then
process these data.

At the reception stage the problem is one of real time, which
we have chosen to solve by recourse to a T 1600 mini-computer as follows :
it consists of a memory of 24 K, of a disc unit, of two magnetic tape
feeders on which the data is stored, of a printer for editing statistics
and of 3 terminals for access to the logic board (2 teleprinters and a
viewing console).

At the present moment intime, only the first half of the processing is carried out in real time on this x T 1600, the second half being

carried out on the IRIS 80 belonging to the Computer Centre in Rennes. At this experimental stage this means that the remote meter readings can be analysed in detail, but the full processing on the T 1600 is foreseeable for the future.

The processing system having been defined in outline, the essential point still to be decided was that of the interface of this system with the network.

In this choice two types of limitation were taken into consideration. The first relates to the design of the TIGRE and to the tree structure of the network, the second concerned the rate determining entry into the mini-computer. Because of the very nature of remote meter-reading, it must be possible to scan the whole system in a matter of seconds. The corresponding flow of information does not make direct access to the memory necessary. However, such direct access to the storage facility could become necessary if the network were to develop towards a pure data transmission service, a development which can be envisaged by virtue of the high bit frequency value selected. Still, in this case, the T 1600 could only be a front, for more complicated equipment would be required to process the data.

Direct memory access having been rejected in favour of the classic style of access through an input-output bus, a decision had to be taken as to the number of couplings to be used. The solution offered by a set made up of one coupling for questions and another for replies was abandoned in favour of a single coupling operating alternately for questions and replies.

For although the first solution offers a higher rate subject to the development of a method for questioning the TIGRE which takes into account their geographical location and avoids a multiple pile-up in the replies, the second offered an input rhythm which is quite sufficient for the present and was even capable of dealing with developments.

The structure of the coupling and the questioning scheme take into account the way in which the TIGRE was developed. In order to simplify the synchronisation of question and reply, a TIGRE being questioned will only reply when it detects the following question. Moreover as the check circuits - for reasons of economy - are common to the question reception and reply transmission functions, it is impossible to direct the following question to the same TIGRE.

The scanning system follows quite naturally from these limitations : question N - reply N-1 - question N+1 - reply N ... Pile-ups are thus rendered impossible and this time-lag avoids any blocking of the system caused by the absence of a reply.

Transmission error detection is made possible by adding redundancy bits in each question or reply. To achieve this, we made use of a systematic Hamming-type cydic code (generating polynome $1 + x + x^5 + x^7$).
Thus can be detected all simple, double, triple and odd-bit errors.

V.2 - Logic Design

The logic performs a set of 17 functions controlled by a real-time monitor (RTES/D).

Certain tasks are guided by the monitor itself, and are thus inaccessible, while others are under the control of the operator.

The logic can be subdivided into 4 groups.

V.2.1 - The CYCLE function : at the heart of the system, it carries out the periodical questioning of all the TIGREs in the network, analyses their reply and stores the information on magnetic tape together with details needed for further use : address of TIGRE, date, time of information.

CYCLE has a 'map' of the network in the form of a disc index which it
consults every time it starts out on its round of questioning (every
minute) and which gives it all necessary background information, viz.

- the address of the TIGRE
- the state of the TIGRE (connected, unconnected, broken down)
- the type of pairs linked to the TIGRE (TV remote invoicing,
  water-meter reading, alarm systems, etc...)
- choice of TIGRE for statistical purposes

Normally CYCLE only consults the index ; but particularly if
it comes across a faulty TIGRE it can introduce an indication of the
changed state of the TIGRE into the index.

It should be noted that the information on the type of TIGRE
will influence the treatment of the data - e.g. water consumption calculation, or transmission of alarms on to a distant terminal.

Finally CYCLE provides, in a common memeory area, available for
other tasks, a report of its explorations.

V.2.2 - The service tasks : specific tasks programmed by the
operator.Amongst these, for example, there may be statistical functions such as measuring the audience level for
the three pay-TV channels, or the editing of data coming
from one or several TIGREs to keep a check on their proper functioning.

Here too we may include the managing of the network map
(disc index) with the ability to add or cut out at any
time a certain number of TIGREs as the network develops.

V.2.3 - Remote supervision function : To ensure the normal functioning of the whole system.

The teleprinter set aside for this task may be installed
any where, e.g. in the office of the central station director or in a
maintenance workshop.

Other possible functions include :

printing out of a CYCLE report
testing the magnetic tape feeders
consultation of network map
testing of a TIGRE

V.2.4 - <u>Other functions</u> : these include the clock and the log
book.

CONCLUSION

At the moment of writing some 10 prototypes of the concentrator
are already or will soon be in communication with the calculator on the
Rennes cable TV network. The experiment is as yet, both in duration and
in size, too limited to allow us to come to any conclusions as to the
reliability of the whole system, but the transmission system is already
operative and giving satisfaction in daily water-meter readings.

A more precise evaluation of the value of this equipment would
be possible once home selection services were introduced.

Cost is certainly an important consideration for its adoption,
but this is difficult to estimate for the adoption of a frequency return
channel does not only imply the installation of a digital transmission
channel, but in addition that of a television channel capable of brin-
ging pictures back from districts of the town to the central station.

Cost also has to be weighed against the advantages it brings :
two-way activity allowing a choice of pictures or facts, or a particular style of teaching at distance, more precise invoicing for taking
certain programmes, and economies in the meter-reading department of
water, gas and electricity boards.

All the possibilities inherent in the system are far from
being taken up, particularly that of a real dialogue between the programme or news distributor and the viewer, who in this way could find
himself personally involved.

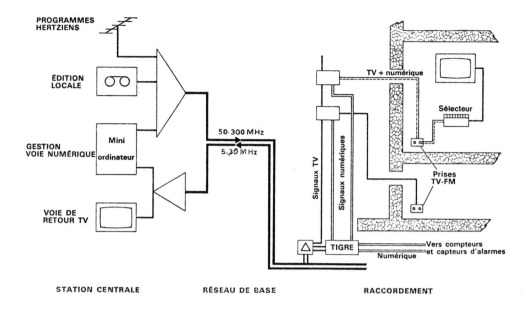

Fig. 1. Lay-out of Two-Way CATV System in Rennes

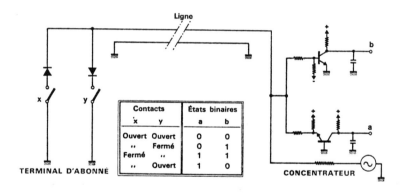

Fig. 2. Principle of Connection between Subscriber and
Tigre Concentrator

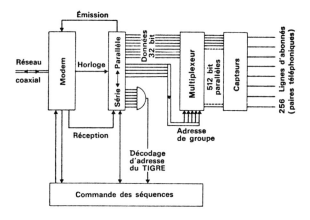

Fig. 3. Block Diagram of Tigre Concentrator

Der digitale Rückkanal im Kabelfernsehversuchsnetz von Rennes

Das für Kabelfernsehnetze mit Kanälen für die Rückrichtung entwickelte System TIGRE erfaßt über das Kabelnetz die Logikzustände von Endeinrichtungen, wie z. B. von Programmwahlschaltern an Fernsehgeräten oder auch von Wasseruhren und verarbeitet diese mit Hilfe eines zentralen Mini-Rechners. Auf diese Weise kann die Einschalthäufigkeit von bestimmten Sendungen und der Zählerstand von Wasseruhren durch Fernablesung festgestellt werden. Jeder Teilnehmer ist mittels einer Fernsprechleitung von max. 500 m Länge an einen der insgesamt 500 im Netz verteilten Konzentratoren angeschlossen. Diese Konzentratoren senden ihre Daten in einem Kanal der Frequenz 7, 64 MHz zurück zur Zentrale und empfangen Daten von dort auf der Frequenz 68, 75 MHz. Eine Erweiterung dieses Systems, z. B. auf den Dialog zwischen Fernsehzuschauer und Programm-gestalter, erscheint möglich.

Cable Distribution in Belgium of Broadcasted Programmes Some of Which Cannot Normally Be Received Locally

Gustave Moreau
Gent, Belgium

Abstract

Today 57% of Belgian television viewers are subscribers of some cable distribution system. Distribution of foreign television programmes, which cannot be received off-the-air, or whose local reception is poor, has come naturally together with the advent of cable networks, as major parts of the country were already very well covered by foreign transmitters often installed close to the national borders. In 1971 the PTT decided to build a national micro-wave system to carry these programmes over the whole of the country and to deliver them to the cable headends. The first PTT link was put into service in the autumn of 1976, but other links existed since 1967. Micro-wave transmission of television programmes does not pose new technical problems. It may be regretted that the same can not be said about the legal aspects of such distribution.

Introduction

Television broadcasts were started in Belgium in the early fifties, but for some years important parts of the country had already been covered by foreign VHF-programmes (mainly the 819 line ORTF programme transmitted from Bouvignies near Lille).
So from the start many Belgian viewers had the benefit of reception of foreign programmes by international spill-over, which in international conferences is often treated as an inevitable but to be regretted accident.
It could be considered there is some hypocrisy in this attitude, when one notes the geographical location of some of the transmitters surrounding Belgium. International spill-over has been the first normal television reception for many Belgians.
When cable networks started their development, main arguments were elimination of unsightly antenna-forrests (= visible aspect of spill-over) and better reception (mainly in towns and valleys), but it should

not be a surprise, that Belgian viewers also took it for granted
that the spill-over situation would be stretched, so that all of the
country could receive what already could be received off-the-air by
half the country. (fig. 2-3-4-5-6).

The history

The first cable television network in Belgium - and as far as I know
in continental Europe - was built by Coditel at Saint-Servais near
Namur in 1961. This first experiment was rapidly followed by Coditel-
networks in the towns of Namur, Liege, Verviers and Visé.
All these towns are mainly situated in valleys, local reception even
of the belgian programmes was poor or very poor and a suitably sited
antenna station was an advantageous solution for the reception problem.
But all these towns are also situated in the french-speaking southern
part of Belgium and only in Namur was off-the-air reception of the
programmes from France possible. Therefore as early as 1967 a micro-
wave link was built from Namur to Liege to bring the at the time 2
ORTF programmes to the main french-speaking town of Belgium. This
micro-wave system was extended to Verviers and Visé, the RTL-programme
was brought to Namur from a reception site at St Hubert, it was included in the existing micro-wave system and it also was brought by micro-
wave to Brussels, where Coditel had in the meantime also started a
cable network.
This was the first phase of cable development in Belgium and it extends from 1961 to 1970. (Fig. 1)
During the years 1968-70 something happened, which made for an"explosion"of cable distribution in Belgium.

On the one side municipal authorities all over the country got interested in the subject and cable distribution became to be considered
as a sort of public service. Amelioration of local reception still
was an argument, but the accent changed to the disappearance of the
antenna-forrests for aesthetic reasons. And these antenna-forrests were
particularly conspicuous in areas of fairly good reception, i.e. in
the flatter northern part of the country, which was rather well
covered by distant or not so distant foreign transmitters. Also the
proportion of one-family houses is still high in Belgian towns and
of course each house had its own antenna-tree, whenever possible
just a little higher than its neighbour's.
Filling in of shadow zones also was an argument, especially in towns
with high buildings. And everybody expected more programmes would
be distributed.

Also during this period other possible advantages were seen, such as
distributing all programmes in the CCIR G standard, which enables
subscribers to use less expensive sets, rather than the luxury multi-
standard set. The advent of colour also put higher requirements on
the quality of reception.
And in some circles local programming became a popular topic.

While there was a general demand, there were on the other side
many offers for building, supervising and managing the new cable networks. Indeed competition was very hard between private companies,
such as Coditel or older radiodistributors, electricity distributors -
private or public - and promotors of ad hoc associations of municipalities.
Most determining factor for the 1970 "explosion" probably has been
the intervention of organisations directly or indirectly linked to
electricity distribution. They had advantages : a suitable administrative and commercial and, within some limits, technical organisation.
And they certainly disliked the idea of seeing the cable network
within their territory built and managed by another similar organisation.
Electricity distribution in Belgium is done for more than 95% by
so-called "Intercommunales", which are associations of municipalities
without ("pure intercommunales") or, in most cases, with a private
partner ("mixed intercommunales").

The result of the struggle is the present situation as illustrated
by Fig.1. Cable networks have been extended from towns to rural
villages and hamlets and the financial situation of some cable companies is not very good. The total amount invested in cable networks
may be estimated at 550 million $ (20.10^9 BF).

Organisation

Cable television is regulated in Belgium by the Royal Decree of
December 24th, 1966 based on art.13 of the Broadcasting Law of
January 26th, 1960, modified on August the 7th, 1961. New legislation
is under discussion. Cable networks are subject to authorization by
the Minister of Posts, Telegraphs and Telephones. The authorization
is given for 18 years and can be prolonged by 9 year periods.
Cable networks must carry the Belgian television broadcasts simultaneously and integrally and may similarly distribute all television

programmes of broadcast stations authorized by the State on which
territory they are established.

Local organisation is forbidden, but some experiments have recently
been authorized in the French-speaking part of the country. Distribution of broadcasts, which have a character of commercial publicity,
are also forbidden. This interdiction, which would force cable
distributors to suppress all publicity in foreign broadcasts, is not
enforced for practical reasons. But this is another and politically
hot problem.

In practice cable distribution is managed in Belgium by :

- private companies : 24% of total number of subscribers ;
- "regies" and pure intercommunales : 36% of subscribers;
- mixed intercommunales : 40% of subscribers.

Mixed intercommunales may vary according to their statutes, which
have to be approved by the Minister of the Interior. They are associations of municipalities with a private partner, which in practice is
the electricity company. They may own their installations or these
may be brought in by the partners for use by the intercommunale. They
may have their own personnel, but nearly always daily management is
provided by the private partner. They have in common that decisions
in the board of directors are taken by qualified majority between
delegates of municipalities and private partner.

The national infrastructure.

As already mentioned the first use in Belgium of a micro-wave link
in relation to cable distribution was made by Coditel in 1967 to
bring the French television programmes from Namur to Liège. It was
extented shortly afterwards and also used to carry the RTL programme.
A second similar system was built by ALE, a pure intercommunale
serving the area around Liège, and this system similarly was extended
by Brutélé, another pure intercommunale, to bring the RTL-programme
to the Brussels area
However all micro-wave is subjected to a licence from the Minister
of PTT, who is responsible for policing frequency use.
And due to the multitude of applications from new cable distributors
and because of the apparent danger, that duplicate or triplicate
systems would be created, the Minister decided in 1971 that no new
licencies would be granted, that the PTT would build a national micro-
wave system and that the taking-over of existing installations would
be negociated in due time.

The national system projected by the PTT and the other existing micro-wave links are illustrated by fig. 7.

The total cost of the PTT system (33 links) can be estimated at 14 million $ (500 million BF), each link being equiped in principle for 4 television programmes in both directions. The PTT estimates the yearly cost per terminal at 70.000 $ (2.5 million BF), basing its estimation on depreciation periods and yearly maintenance costs of respectively 50 years and 5% for buildings, 20 years and 2% for towers and 12 years and 5% for the equipment. Interest rate was taken at 8,5%.

The first intention of the PTT was to limit the national system to the delivery of programmes to the main towns or cable headends (as indicated on fig.7) and the following rates were proposed : 40 BF/ subscriber/year for one programme, 27 BF for a second programme, 23 BF for a third programme and 20 BF for each following programme. These rates, which followed after a previous higher proposal, raised many objections from the cable operators, who considered them to be still not only rather high, but also favouring the main towns or headends , as further distribution by means of multidirectional one-way micro-wave would have to be paid as a supplement or cable opera-tors would have to invest themselves in cable links to the RTT terminals (supertrunks), what some in fact already had done. They did not feel the treatment was equitable. As a result the PTT last year put forward new rates, which finally were accepted. These final rates include further distribution by the PTT to all cable headends and they amount to 60 BF/subscriber/year for one programme, 40 BF for the second, 30 BF for the third and 20 BF for each following programme. In due time existing supertrunk links from the cable headends to the PTT terminals would be taken over by the PTT, but in the meantime the previous lower rates will apply for cable operators having themselves made the necessary investments.

The PTT link Brussels-Courtrai-Ghent is in operation since september 1976 and its extension Courtrai-Adinkerke-Bruges should be put into operation next June. The following programmes are carried : RTL, ARD1 and ZDF.
Micro-wave links are a well - known and proven technique, so there do not seem to be technical problems worth mentioning.
However at present, due to reception conditions of the German programmes in Brussels, quality of these programmes is not too good.

The Local infrastructure.

We take as example two mixed intercommunales : Teveoost and Tevewest, both having the same private partner. The Teveoost network covers the area around the town of Ghent including 103 small towns and villages, but without the Ghent agglomeration, which is served by another mixed intercommunale, Tevegenta (50.000 subscribers). It has 3 headends, 1900 Km (1184 miles) of cables, 43.700 subscribers on a potential of 87.700 (49.8% penetration). 12 television programmes are distributed: BRT1, BRT2*, RTB1, RTB2*, NOS1, NOS2, TF1, Ant2*, FR3*, RTL, ARD1* and ZDF* (*with set top convertor). The 3 last programmes are supplied by the PTT micro-wave system at the Tevegenta headend in Ghent.

The Tevewest network covers Bruges and the area around Bruges excluding the town of Ostend. Ostend is an older concession of the private partner, but part of the Tevewest area is supplied from the Ostend headend. Tevewest has 3 headends, not including Ostend, 1820 Km (1129 miles) of cables, 60.000 subscribers on a potential of 115.000 (52.2% penetration). 15 television programmes are distributed : BRT1, BRT2*, RTB1, RTB2*, NOS1, NOS2, TF1, Ant2*, FR3*, BBC1*, BBC2, ITV* and, starting from next june, RTL*, ARD1* and ZDF* (*with set top convertor).

In both Teveoost and Tevewest all television programmes are distributed in the PAL-standard and both also carry 15FM-programmes. As illustrated by fig. 8 all headends are connected to the centrally situated headends of respectively Ghent and Bruges, which are terminals of the national PTT micro-wave system, by means of low-loss, dual cable underground links (supertrunks) using amplifiers having a frequency range of 30 to 120 MHz. Each cable at present has a capacity of 6 television programmes using alternate channels. Alternate channels were chosen because we had some - unjustified - fears of cross-talk between both cables.

The decision to build the supertrunk systems was taken to be independent of PTT timing after the PTT had decided not to grant new frequencies for micro-wave links.   A comparison between 20 Km cable links and micro-wave had shown the supertrunk to be an economically sound solution and it gave us the certainty, that all headends would be served simultaneously, when the national PTT system would be ready in Gnent and in Bruges. Also taken into account was the fact that for over 75% of their length the supertrunk cables could be laid in the same trench and together with primary feeders, which we also had decided to lay underground.

The supertrunks have allowed us to concentrate in the Ghent and
Bruges headends all the equipment for transcoding the French SECAM-
programmes to the PAL-standard (as well as, up to last Christmas, for
translineating the French 819 line programme to 625 lines CCIR). Also,
in the case of Tevewest, the supertrunks allowed to bring the British
programmes to Bruges, where urbanistic regulations were a difficulty
for erection of the necessary antennas. Not including the terminal
equipment at the headends, the total marginal cost of the supertrunk
system (64Km / 40 miles in Teveoost and 79Km/49miles in Tevewest)
can be estimated at 1.7 million $ (62.6 million BF) including 5% VAT
and 15% design and supervision costs.
The performance of the supertrunks is checked **monthly** during headend
supervision and a more extensive check is foreseen every 6 months.
Over a 12 month period the supertrunk system has required a total
of 26 trouble interventions, 9 of which during evening or weekend
hours.
As an overall appreciation of the Teveoost (system B) and Tevewest
(system A) cable distribution systems fig. 9 and 10 give a comparison
with figures, which were published by TVC-magazine (Dec 76) and which
give the results of an inquiry in US cable systems.

The Bern copyright convention and Belgian jurisprudence.

As we have seen distribution of programmes, which cannot be received
locally, does not pose special technical problems to the cable distributors. It is not te be contested that such programmes are a supplementary incentive for people to become subscribers : cable systems are
expected to extend the choice of programmes. Whether these programmes,
which in a small country like Belgium of course are foreign programmes,
are actually viewed and, if so, which parts of them, is difficult to
say at the moment as we do not yet have valid viewing statistics. One
can however say that the language of the programme certainly is a
factor and may be a barrier, at least for a major part of the population. It is very normal that the first micro-wave links were built
to bring French programmes to French-speaking towns and they did not
generate international cries of piracy or pilfering. Unfortunately
the same may not be said about the present situation. In fact this
situation is unclear partly because copyright holders in recent
years have refused to distinguish between cable distribution of broadcasted programmes received off-the-air and distribution of broadcasts,
which are not or can not be received locally.
It is in this context and under pressure from our subscribers and
from our public partners that we have taken the risk of distributing

programmes brought to us by the PTT micro-wave system without legal
cover or with little cover.
The discussions all go back to the interpretation of article 11bis
of the **Berne** copyright Convention as revised in Brussels in 1948.
Famous art.11 bis reads as follows :

"(1) Authors of literary and artistic works shall have the exclusive"
"   right of authorising : 1°) the radiodiffusion of their works   "
"   or the communication thereof to the public by any means of   "
"   wireless diffusion of signs, sounds or images; 2°) any com-   "
"   munication to the public, whether over wires or not, of the   "
"   radio-diffusion of the work, when this communication is made   "
"   by a body other than the original one; 3°) the communication   "
"   to the public by loudspeaker or any other similar instrument   "
"   transmitting, by signs, sounds or images, the radiodiffusion   "
"   of the work.   "
"(2) It shall be a matter for legislation in the countries of the   "
"   Union to determine the conditions under which the rights   "
"   mentioned in the preceding paragraph may be exercised...a.s.o. "

After 1948 and due to this article radiodistributors in Belgium paid
an indemnity (about 3%) to the Belgian authors organisation (SABAM),
but all these radiodistributors also distributed a locally originated
programme, they made a choice of stations to be distributed and distributed them simultaneously but successively and their subscribers did
not need to have reception equipment, which connected to an antenna
would have enabled them to receive the same programmes. (They did
however pay normal radiotaxes). The agreement with SABAM gave a good
cover for SABAM represented practically the entire repertory used in
sound broadcasting.
When Coditel started cable television distribution in 1961 the SABAM,
after discussion, agreed that, this activity being similar to the
service rendered by a private or collective antenna, no copyright had
to be paid. Only in 1967, when Coditel started retransmitting French
programmes by micro-wave, an agreement was made, whereby Coditel
paid a copyright due to SABAM for such use of the French repertory
used in these broadcasts and represented in Belgium by SABAM.
Since then the importance of cable television distribution has grown,
the position of SABAM and indeed of most other copyright holders has
changed and they now request copyright liability for all distribution
of broadcasted programmes by cable systems. This is so nationally
and this is so internationally.

Cable distributors obviously can not accept this (for it would be
an _exclusive_ right), but nevertheless and after several years of
negociation a compromise was reached between SABAM and the national
cable distribution association (RTD), whereby, pending jurisprudence
by the Belgian courts, cable distributors will pay a copyright due
of 1,5% of mean annual subscription revenue only when broadcasts are
distributed which are brought in by micro-wave or by supertrunk cables.
It is to be regretted that SABAM only represents a fraction of copy-
righted material used in television broadcasts, and there are othher
raiders before the coast.

How does Belgian jurisprudence stand as regards copyright liability of
cable distribution ? In 1969 the Brussels Appeal Court in a case opposing SABAM to a radiodistributor stated that copyright had te be
paid, because the distributor "does not limit himself to retransmitting the Belgian and foreign transmissions chosen by him, but that he
receives, filters, transforms and amplifies them so as to assure his
subscribers of a continuous reception free of interference; that he
acts independently; that his activity must be considered as a performance in itself and as a public communication next to and seperate
from the properly so called radiophonic transmission" (free translation !). One may wonder wonder what this precisely means !

More important is the pending case opposing CINEVOG to Coditel. Judgment
in this case, which is under appeal and which forms the jurisprudence
to which the SABAM agreement makes reference, was pronounced in Brussels
in 1975. The subject was the distribution by Coditel of a French film,
which had been broadcasted by German television, while this film was
still running in Belgian cinemas. Coditel receives the German programmes by air, so the Coditel-SABAM agreement was in no way involved.
Cited together with Coditel was the French producer, who had given
exclusive distribution rights in Belgium to CINEVOG and who had given
television rights to German television, although he should have known,
that the German broadcasts are received in a major part of Belgium. He went
however free for lack of evidence of his responsibility. Coditel was
condemned to pay CINEVOG 300.000 BF for material and moral damages.
The argumentation is however interesting. The Court referred to art.
11 bis and examined whether there had been a communication, whether
it had been public and wether it had been made by another organisation. The Court stated that art 11 bis (1) 2°) aimed "under certain
conditions the public communication of a broadcasted spectacle to a
new circle of spectators". It estimated that there is a communication
in so far as that there is an active intervention prior to reception
by the viewers and intended to allow this reception. The intervention

by Coditel had been not only real but important, making possible by various technical means (cleaning, eliminating interference, amplifying) a reception, which would have been non-existant without this activity.

As regards the public character of the communication, the Court stated that it was public by destination in so far that it was accessible to all prepared to pay the required price, thus trying to distinguish a cable system from a collective antenna. This clearly is giving a different interpretation to the term "public" as the signification given normally to the same term in the next phrase of art 11bis : a communication by loudspeaker is considered public when the loudspeaker is installed in a place accessible to the public. Finally the Court decided that the activity of the distributor was of such importance as to be distinguished from the broadcaster's activity, so that the criterion of a body other than the original also should apply.

Several of the argumentations of the Court may and have been questioned and this illustrates the confusion, which exists in the interpretation of the terms used in art 11 bis and which have not been precisely defined in the Convention.

The judgment in fact does give no clear answer to the question whether cable distribution of directly received programmes is liable to copyright or not. Indeed the distribution of German programmes for instance in Brussels can be considered to be a border case, where, due to distance and geography, only suitably situated individual or collective antennas could receive the programmes. Overlooked also is the fact, that, if cable distribution within the direct reception zone were subject to the exclusive right of authorization, this would also have to apply to distribution of the national programmes, and this would be in opposition to the legal obligation in Belgium to distribute these programmes integrally.

When one reads articles on the legal problems of cable distribution, it is a striking fact, that most authors come to the conclusion that cable systems should be liable to copyright, but that collective antennas should not. To come to this conclusion they all have to struggle with the interpretation of the terms used in art. 11 bis and for a non-legal mind some of the argumentation is rather peculiar. In German articles one often tries to detect the existence in cable systems of some sort of transmitter, which would not be existent in even large collective antenna-systems. For a technician this of course is nonsense.

In a French article I read that a loudspeaker is an out-moded piece
of equipment now usually only found in public spaces and that a
micro-wave link could be called a transmission by wire ("câble hert-
zien"), while a coaxial cable could be termed a wireless transmission,
the cable being only the envelope wrapping the hertzian waves.
All this to make national legislation consecutive to art. 11 bis
mean cable distribution of broadcasts to be liable to copyright.

In the report of the Brussels convention (1948) one can find that the
originally proposed wording of art. 11 bis (1) 2°) was : "any new
communication to the public whether by wire or not of the radiodiffusion of the work" and it is indicated that this did not mean wire
or cable distribution within the reception zone. The proposed text
was changed as it stands now, because it was unacceptable for the
broadcasters. Could this modification have been intended to include
all cable distribution without mentioning this in the report ?

We think it would be better to stop being obsessed with the interpretation of a thirty-year-old text and to try to find a logical,
equitable and workable solution for the real problems posed by cable
distribution systems. For all the previous considerations leave us
out in the cold when distributing broadcasted programmes simultaneously
and integrally outside of the direct reception zone. And this in
our opinion is the real problem for which a workable solution has to
be found. It must be a workable solution, for it should be realised
that the cable distributor in this case is in the practical impossibi-
lity to obtain, prior to distribution, all the necessary authorizations
from the vast number of right-holders, which moreover are unknown to
him. Will there be another solution than some sort of compulsory
licence ? We might imagine an agreement with the concerned broadcasters,
but this may also only displace the problem. In any case black-outs
should be avoided, for they would be a nuisance for the cable distributor and would surely not be appreciated by the subscribers.

Rome Convention on neigbouring rights (1961)

This convention covers the protection of the rights of performers,
producers of phonograms and broadcasting organisations. It is up
to now of less importance for cable distribution as this distribution
is not specifically mentioned in the convention text.
More important however is the European Agreement on the protection
of television broadcasts (Strasbourg 1960).

According to art 1 - 1(b) of this agreement broadcasters may prohibit
diffusion to the public by cable of their programmes, although
Parties (i.e. countries) to the agreement may withhold this protection.
This possibility of withholding protection has been reduced in 1965
to a percentage of the transmissions of not more than 50% of their
average weekly duration. Since 1975 and due to a proposal by the
European Broadcasting Union the legal committee on broadcasting and
television of the European Council has discussed the opportunity of
suppressing the reservation alltogether while introducing the notion
of the direct reception zone, within which cable distribution of the
broadcast would not be protected. The direct reception zone was defined
as the area, within which the signals emitted can be received at an
acceptable level of quality by means of reception facilities normally
at a private viewer's disposal. An alternative proposition limited this
zone to the less realistic concept of the service area as defined by
CCIR parameters.

No agreement was reached, most opposition coming from the bigger
countries, where of course reception of foreign programmes is relatively limited to border areas. The bitterest opposition came however
from the German delegation. During last january's meeting it was decided,
due to the possible implications on copyright and neighbouring rights,
to wait for the results of discussions to be held within the framework
of the Berne and Rome conventions and which are scheduled to start
this year in june. These discussions will be continued on different
levels in 1978 and probably 1979.
One of the main objections against the concept of the direct reception
zone was that the definition was not sufficiently precise and that
the zone could extend in the future with the development of reception
technology and with the capacity of cable system to provide for more
expensive and better performing receiving equipments.
Personally I think the opposite will happen, provided a decent solution can be found for the indemnisation of rightholders when distributing broadcasts outside the direct reception zone. And so the problem
of the delimitation of the direct reception zone might become a false
problem. Indeed, due to the quality cable distributors want to provide
to their subscribers and which the subscribers expect and request,
cable systems will tend to import broadcasts rather than, by means of
elaborate reception equipment and with more or less success, try to
stretch the direct reception zone beyond its reasonable limits. This
anyway is the Belgian experience.

Conclusion

Cable distribution of broadcasted programmes, which can not be received locally, is an activity, which cable subscribers in Belgium
take for granted.
Technically there are no specific problems, but legally, both on the
national and on the international level, this activity raises
difficulties for which there is no practicable solution at present.
We sincerely hope that such a solution will be found in the near
future an that it will be realistic and equitable. The SABAM-agreement
may have been a first step.

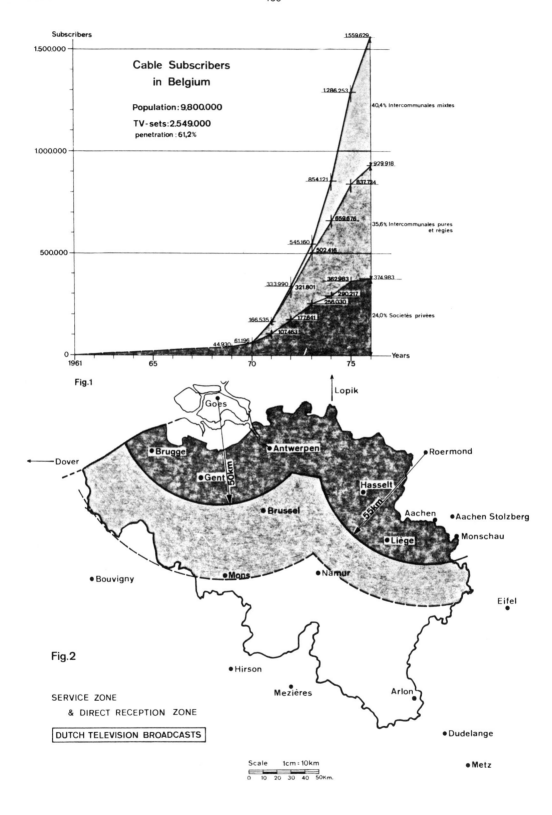

Fig.1

Cable Subscribers
in Belgium

Population: 9.800.000

TV-sets: 2.549.000
penetration: 61,2%

Fig.2

SERVICE ZONE
& DIRECT RECEPTION ZONE

DUTCH TELEVISION BROADCASTS

Scale   1cm : 10km

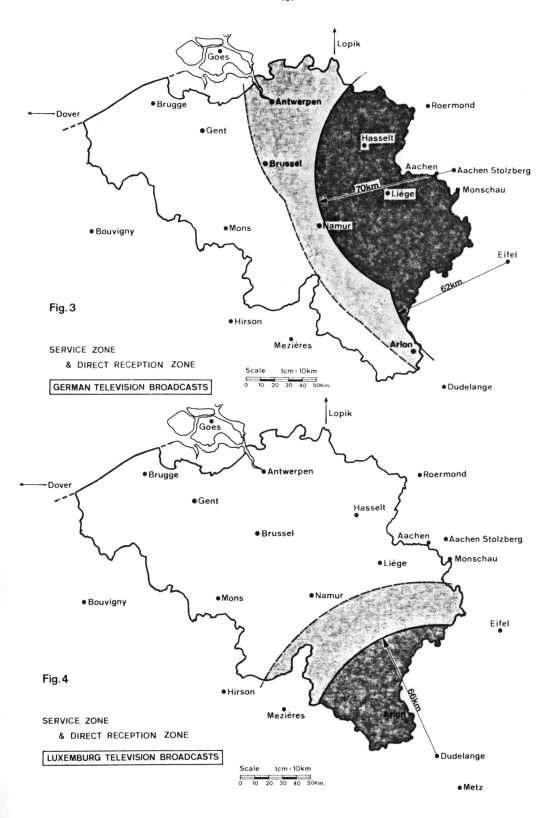

Fig. 3

SERVICE ZONE
& DIRECT RECEPTION ZONE

GERMAN TELEVISION BROADCASTS

Scale   1cm : 10km
0   10   20   30   40   50Km.

Fig. 4

SERVICE ZONE
& DIRECT RECEPTION ZONE

LUXEMBURG TELEVISION BROADCASTS

Scale   1cm : 10km
0   10   20   30   40   50Km.

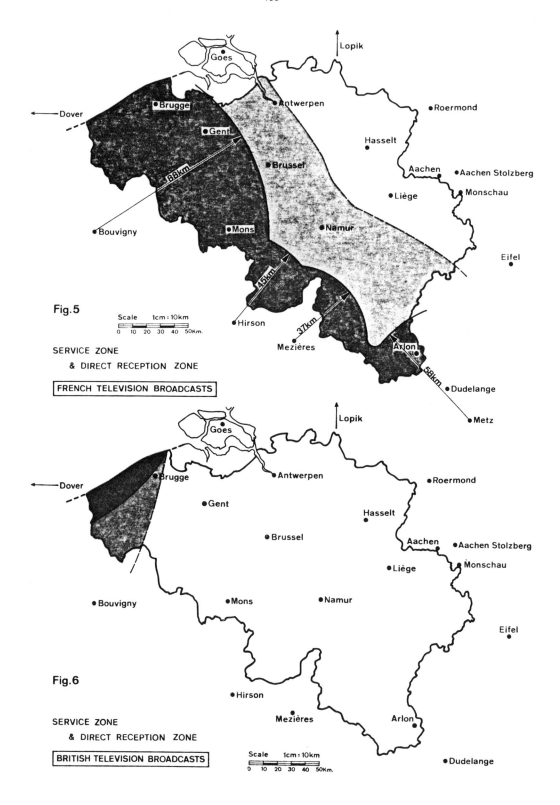

Fig.5

Scale 1cm:10km
0 10 20 30 40 50Km.

SERVICE ZONE
& DIRECT RECEPTION ZONE

FRENCH TELEVISION BROADCASTS

Fig.6

SERVICE ZONE
& DIRECT RECEPTION ZONE

BRITISH TELEVISION BROADCASTS

Scale 1cm:10km
0 10 20 30 40 50Km.

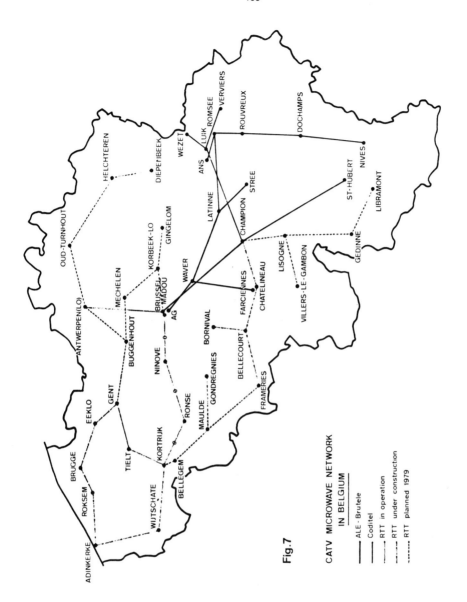

Fig.7

CATV MICROWAVE NETWORK
IN BELGIUM

ALE - Brutele
Coditel
RTT in operation
RTT under construction
RTT planned 1979

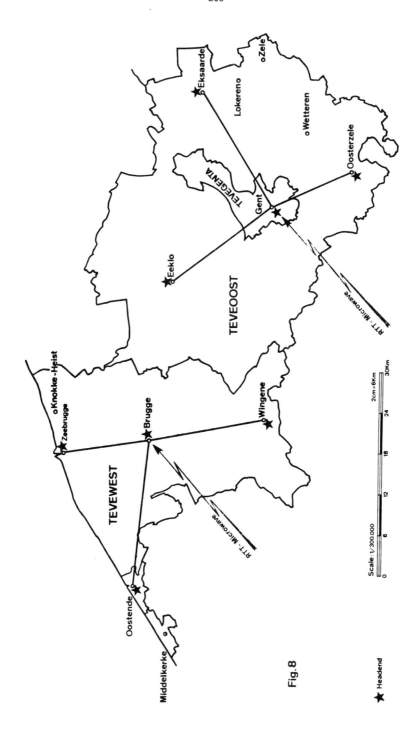

Fig. 8

from TVC Dec.1976

TABLE I				Systems		
	High	Avg	Low	A	B	
T.C./1000 Subs/WK	18,8	8,0	1,4	0,87 1,18 1,48	1,98 2,67 3,82	Without converters With 100% converters
Maint Hrs/1000 Subs/WK	28,0	16,1	10,0	4,91	8,21	
Subscribers/Tech	2800	2109	300	5230	2918	
Miles/Tech	110	47	20	101	103	
Est.Cost/T.C.	$40	$14	$4	$29	$19	

System A: 3 headends
1129 miles
58.421 subscribers
50,7% set top converters
50,8% penetration

System B: 3 headends
1184 miles
36.385 subscribers
37,1% set top converters
43,3% penetration

(1 US $=36,885 BF)

Fig.9

01.77

from TVC Dec.1976

TABLE II		Systems	
	% of Total Trouble Calls	A	B
Set or Fine Tune	43%	12%	9%
Drop	} 28%	22%	17%
Converter		26%	26%
System	14%	15%	28%
Miscellaneous	15%	25%	20%

With 100% converters the converter trouble call % would be:

A	B
41%	48%

Fig.10

01.77

Kabelübertragung von Rundfunkprogrammen in Belgien, von denen einige unter normalen Bedingungen örtlich nicht empfangen werden können

Gegenwärtig sind 57 % des belgischen Fernsehpublikums an Kabelverteilsysteme angeschlossen. Die Verteilung von ausländischen Fernsehprogrammen, die entweder überhaupt nicht oder nur schlecht über die Antenne empfangen werden können, bot sich mit der Einführung von Kabelnetzen geradezu an, insbesondere, da große Teile des Landes bereits in Reichweite der teils in der Nähe der Landesgrenzen installierten ausländischen Fernsehsender liegen. Im Jahre 1971 beschloß die PTT die Errichtung eines nationalen Mikrowellennetzes, um diese Programme über das Land übertragen und den verschiedenen Kabelfernsehzentralen zuführen zu können. Die erste Übertragungsstrecke der PTT wurde im Herbst 1976 in Betrieb genommen. Andere Übertragungswege gibt es schon seit 1967.

Die Übertragung von Fernsehprogrammen über Mikrowellennetze wirft keine neuen technischen Probleme mehr auf. Leider kann man dies von der rechtlichen Lage nicht ebenfalls behaupten.

A CATV-HF System Experimented in Italy With the Possibility of Backward Channel

Luigi Bonavoglia
Roma, Italy

Abstract

An HF-CATV system developed and experimented recently in Italy is described; important features of the a.s. system are the attempt to make as much use as possible of the facilities offered by the ordinary telephone network and the study of the possibility of integration of CATV with other types of communication such as data transmission, video-phone, wire-sound broadcasting and of course telephony.

The system is of the star type and selection of the desired program is made by users through a dialling set which by remote control connects the user's line to the wanted program.

Possibilities offered by this system for the installation of backward channels are examined.

General

Diffusion of information from one central point towards many users has been a problem which goes back to the invention of telegraphy on metallic wires; the appearance of radio broadcasting was considered to be the appropriate solution of the problem. However, as the amount of information has increased, especially passing from audio to TV broadcasting the possibilities offered by transmission on wires, particularly on cables, have been considered very attractive.

It is common today to consider the plant which was installed in 1949 in a western mountain region of the USA as the first application of CATV. From that time CATV has made rapid progress until in these last years another need has arisen, that is to collect information with backward channels from the users, i.e. from the periphery to the centre of distribution. We will discuss this second part of the problem at the end of the paper: at the moment we want to point out that the difficulties met in this second part of the problem are quite different from those which arise in the diffusion from a central point to the periphery.

Firstly, we will report on the results obtained in Italy by the use of a CATV system which for its particularities can be rather easily employed for the implementation of backward channels.

This system has been described in some detail in different papers presented at international congresses; therefore, in this paper we will limit ourselves to a short description and to a report on the results of the experiment.

CATV-HF System

The main idea on which the project was based at the time of its original study was that of taking the maximum advantage of the facilities offered by the existing telephone network.

If we examine a CATV network, three parts can be identified:
a) customer lines linking users to some nodal centres;
b) links from nodal centres to the point where programs are put in the form apt for distribution; (in the following we will call this point head-end);
c) links from head-end to studios generating or receiving programs.
It is well known that for the part under a) different possibilities exist and since we do not want to enter into an analysis of this variety of solutions, we will pass directly to the description of the system studied by SIP (Società Italiana per l'Esercizio Telefonico S.p.A.) and developed with the assistance of some Italian manufacturing companies.

The solution selected for this system is such that TV signals can be sent to users by means of the ordinary telephone lines which are usually twisted pairs in underground or suspended cables.

It was decided for this reason to use a distribution band low enough (in HF band) to be accepted on these pairs and high enough that in the band under the TV signal other services could coexist including the telephone service, data transmission, wire-sound broadcasting (this last service exists in Italy from a rather long time). The choice of frequency allocation and modulation system is described in Fig. 1. We must only point out that variations could occur in the future due to the decision that will be taken on the video-telephone: in this respect the world situation is evolving and changes in the band-width of this service, which will influence the frequency allocation of our system, can be expected.

The 13.2 MHz frequency at the top of the band is used for remote controls coming from the user. It can be noticed that the TV signal is single band modulated with vestigial and allocated between 2.5 and 9.25 MHz; a simple frequency reallocation permits reception by ordinary television sets.

Figs. 2 and 3 show the general organization of the broad-band distribution system in which the following parts can be identified:
a) Head-end
b) Trunk line
c) Nodal centre (or switching box or switching centre)
d) User's distribution network
e) User's terminal

a) The head-end (Fig. 4) consists of two sections: one for the production or reception of programs, the other for modulation and translation of signals in the line band. The latter provides for two separate inputs for video and audio signals. The video signal is amplitude modulated and transposed in the line band, while the audio signal is frequency modulated and added to the video signal. All the transmitting equipment of the terminal is of high level standard ensuring to users an image quality at least equivalent to that which can be obtained off-air in a good reception area.

b) The trunk line is arranged with space division by means of amplified lines in microcoaxial cable which is a type of cable normally used in Italy for transmission of medium capacity telephone channels. Two types of repeaters are envisaged: one for connecting lines and the other for distribution lines. The former (Fig. 4) is for connections between the head-end and the switching centres and the latter for the feeding of the switching matrixes. Economy and reliability considerations make the use of 2 Km spaced amplifiers advisable. For the extension of the system range, 1 Km spaced amplifiers can be used. In this way distances of up to some tens of Kms can be reached.

c) The signal forming the programs already transposed at the same frequency allocation employed by the user is transmitted over micro-coaxial cables and is concentrated at the nodal centres. The characterizing element of the nodal centre is the switching matrix (Fig. 5) sited in a telephone type cabinet and fed locally. The switching matrix permits the user to select the desired program by means of a suitable remote control, using the 13,2 MHz signal shown before. Besides the switching unit, the switching part contains the branching filters which permit the insertion of the signals of the other services of the integrated network on to the user's line. A matrix of modular structure can be installed in order to simplify the addition of subsequent extensions.

d) The user's distribution network connects the switching matrix with the user. Ordinary telephone cables with twisted balanced pairs can be used in some cases for a limited percentage of the number of pairs contained in the cable. Special telephone type with screened pairs can be used and in this case the transmission of a television program together with all the other telecommunications signals can be carried on each pair.

e) The user's distribution terminal consists of the following parts:
- branching filters for the signals required by the user and for protection from the line;
- interface equipment for the television set, consisting of the HF-VHF converter and the dialling program set (Fig. 6).

After this short description of the system some considerations about the various parts should be added.

Trunk Lines

The use of microcoaxial cables (0, 7 - 2, 9 mm) has been foreseen because this type of cable is going to be used very extensively in the Italian telephone network, both for medium distance connections, as in local areas for connections between switching offices. Normally on this type of cable a 34 Mb/s PCM system is obtained with a spacing of 2 Kms for the regenerating amplifiers.

This situation suggests adopting the same spacing of 2 Kms for TV transmission in order to utilize the same containers as those used for telephone service. Being the TV signal allocated rather high in frequency, no crosstalk problems arise, taking into account the fact that the cross-talk ratio, which is rather low under 100 KHz, rapidly increases with frequency in this type of cable. Amplifiers are fed from one terminal and are automatically gain controlled by using as reference the synchronizing peak of the TV signal. Protection against surges, essentially of the same type used in telephone carrier systems is provided.

Nodal Centres

These centres are characterized mainly by the existence of a switching matrix, which has the task of connecting to a certain program all the users who wish to receive it. The matrix has been object of careful studies with the aim of obtaining high reliability, as little maintenance work as possible and a size which could allow its installation inside the existing cabinets of the telephone network ordinarily used for permutation of pairs.

The switching element finally chosen is a solid state component especially designed to meet all the requirements necessary to maintain the desired high standard quality of the image. Through attenuation, frequency response and crosstalk have been kept within the desired limits using a modular matrix: each modulus of the matrix makes use of an integrated circuit which is capable of accepting 8 programs on one side and of feeding the user's line on the other. The number of integrated circuits per modulus must be fixed in order to achieve a good economic balance: it appears that a modulus of 8 programs x 20 users should be a sensible solution; amplification on both sides of the matrix is needed and included in the modulus. Each integrated circuit contains also a decoding circuit which receives from the user the 13.2 MHz signal for the selection of the program. Being this matrix the most delicate point in the system it may be useful to produce the measured performances in a field trial carried out in Rome.

Attenuation:	0 dB $\pm$ 1
Bandwidth:	1 - 20 MHz within $\pm$ 0, 75 dB
Harmonic distortion:	2^ order < 0, 1% (i. e. 60 dB)
	3^ order < 0, 05% (i. e. 66 dB)
Overall Crosstalk:	> 55 dB

User's Distribution Network

It has already been said that it was considered an important economic point to utilize the ordinary telephone plant as much as possible for this part of the system; therefore a careful examination was conducted especially to decide if telephone pairs were capable of transmitting the TV-HF signal without producing an intolerable degradation of the quality.

Very soon it was cleared that limiting factors were level of noise in the cable and far-end crosstalk. Attenuation and impedance regularity were considered of good level. The measurements on many existing cables laid in towns revealed that HF-CATV transmission was possible on a certain number of pairs in each cable; cables with pairs divided in groups offered more possibilities than the others. It was considered useful, however, to study special cables with screened pairs in order to overcome the limitation in number of useful pairs available in normal cables.

Having at this moment a good knowledge of the possibilities offered by ordinary pairs and by screened pairs a philosophy was established concerning the installation of CATV-HF systems in the case of superposition to an existing telephone plant, and in the case of contemporary installation of CATV and telephone plant. In both cases for CATV full use is made of facilities like underground ducts, underground housings, cabinets etc. Obviously better arrangements can be obtained in the second case.

User's Terminals

The user's terminal consists of an ordinary television set, an HF-VHF converter, branching filters for separating the TV signal from the other signals on the line.

Furthermore a push-button control set is provided; this set sends to the switching matrix, by manipulating the 13, 2 MHz carrier with proper codes, the necessary information of the program to which to connect the user's line. The 13, 2 MHz carrier is sent over the same pairs as the other signals.

Field Trial of the System

After the development work, each part of the system was tested in partial experiments; having judged satisfactory the results obtained, an experiment using equipment manufactured by various companies was set up in Rome: the HF-CATV system built up for the experiment transmitted eight programs obtained by reception of the two national TV programs, from a TV camera, from a video recorder and from bar generators, or test-pattern generators. Programs were in colour and in black and white.

From the head-end it was possible, in the experiment to send the 8 programs on amplified microcoaxial lines, from these to feed one 8 programs x 20 users matrix and via a screened pairs cable to feed 20 users. The experiment was successful: the feasibility of building up a matrix with some hundreds users and about 20 programs was demonstrated by the technical results.

Backward-channels

It is quite clear that a point to point channel, fixed or switched, can always be associated with a channel in the opposite direction without any philosophical difficulty. Telephone networks are remarkable examples of this way of acting; in the beginning of this art and also now, for some systems the go and return channels were (or are) on the same physical support. When we look at the distribution of the same information from one point to many users what is meant by return or backward channel becomes something which must be defined specifically because the kind and amount of return information we are speaking of is not obvious. The first point we have to consider is the number of possible sources (among all the users); secondly, the amount and type of information they have to send and finally who or what machine must receive this information.

To clear these points let us give an example: let it be a star network of bidirectional TV channels, in which all the channels from the centre are fed by the same signal, for instance an educational one; the users in this case can individually transmit to the centre questions and ask for explanations, but if only one teacher is present they must use their facilities one at a time. Therefore for most of the time, the individual return channels are idle resulting in a rather non-economical conception of the plant. One can go on discussing the infinity of situations which can be imagined but the main limiting point will always appear to be the capability of the centre of absorbing information from the periphery and treating it; once defined this capacity, for instance, in terms of bits, kbits or Megabits per second, it will be more defined the type and the number of return channels which can be associated to the distribution without offending not only the economy but also common sense will.

Actually a solution giving to the total capacity of the return channels an amount equal or slightly higher than the capacity of acceptance of the centre appears a good one: from a trend of this type, for instance, a solution of many narrow channels (maybe as many as the users) or a solution with few broad channels connecting some selected pick up points (fixed or switched) to the centre may be derived.

Once made this choice i.e. if the backward channels must be a few with wide-band or many with narrow band, it can be decided the best technical way of accomplishing the desired task.

The system already described can be adapted to both these extreme possibilities: in the case of narrow backward channels an auxiliary carrier can be added on the top of the used spectrum and used by the customers in order to give information to the centre. At the nodal centres a logic unit can be added, processing the information sent by the users coding the auxiliary carrier; by the nodal centres information after processing is sent to the head-end. Examples of services obtainable in this way are:

- listening index: the head-end can obtain almost automatically the amount of listeners to each program.
- question answering: if the questions are put in a simple form requiring only yes or no answers. In this way market inquiries can be easily carried out. Answers can be summarized or can maintain their individuality.
- user's exclusion: particular channels can be excluded to some users on control by the head-end.
- pay TV: the selection by the users of some TV channels can be recorded in order to be taxed according to special contracts.

Of course many other possibilities of services arise obtainable by the use of narrow band backward channels. A detailed analysis of the technical problems involved is reported in reference n. 17.

Some words must finally be said on the possibility for the HV-CATV System of providing few wide-band backward channels. The most natural solution is that of selecting some spots as pick-up points and to use another twisted pair from these spots to nodal centres in order to build up a wide band channel which can be a TV one.

A second switching matrix may be installed in some nodal centres and the general arrangement of the system is shown in Fig. 7. The use of matrices also in the backward direction gives the opportunity of using only a limited number of trunk lines for a higher number of possible pick-up points. The control of the matrix comes in this case by the head-end which can decide which pick-up point is to be connected.

The study of the technical problems connected to this system is in progress and it appears that a solution like the present one, using screened pairs, is a first step towards a higher integration of the services of many different types that in the future can be offered to customers.

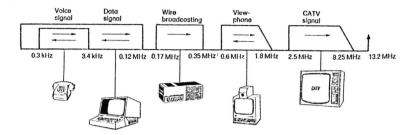

Fig. 1 - Allocation of signals simultaneously present on a user's twisted pair.

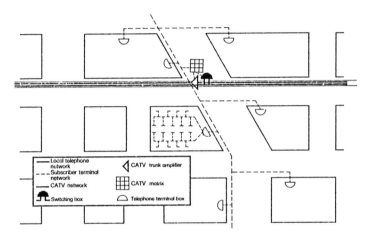

Fig. 2 - CATV distribution network

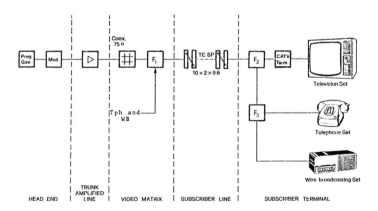

Fig. 3 - Block diagram of the CATV system

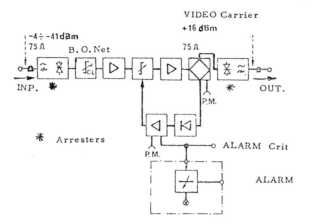

Fig. 4 - Repeater

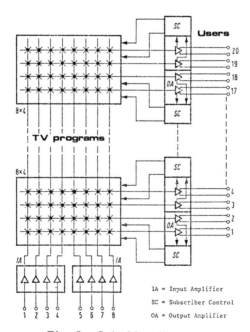

IA = Input Amplifier

SC = Subscriber Control

OA = Output Amplifier

Fig. 5 - Switching diagram

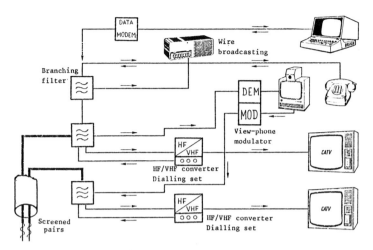

Fig. 6 - Subscriber terminals

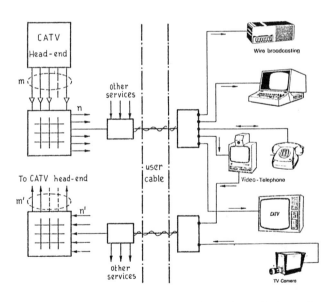

Fig. 7 - HF-CATV System with backward channel

BIBLIOGRAPHY

1. Paladin G., Savino A.: Modelli di rete di distribuzione in cavo e relativi sistemi per fornire servizi di telecomunicazioni. (a)

2. Bassani G., Lauro Grotto U., Marchetti G.F.: Commutazione video e distribuzione di programmi relativi su coppia simmetrica. (a)

3. Gualano S., Rizzo F., Milanese L.: Problemi di trasmissione del segnale video su cavi coassiali e su cavi a coppie simmetriche ed apparecchiature relative. (a)

4. Morganti G., Santoro G., Silveri G.: Studio dei problemi di qualità di un sistema di CATV al fine di pervenire ad una ipotesi di ripartizione di tolleranze. (a)

5. Bellato L., Cichett M., Vannucchi G.: Problemi di trasmissione nella distribuzione di segnali televisivi via cavo in banda HF. (a)

6. Kay A.A.: Technical, operational and economical problems of H.F. receivers, commercial type. (b)

7. Morandi G., Squadroni A.: Adattatore per ricezione in HF. (b)

8. Morganti G.: Problemi generali di qualità per le reti di CATV. (a)

9. Morandi G., Squadroni A.: Obiettivi di qualità in un sistema di distribuzione CATV di tipo a stella con commutazione. (a)

10. Bucciarelli T., Picardi G.: Trasmissione di segnali televisivi su un cavo microassiale. (c)

11. Milanese L.: Realizzazioni costruttive e misure sperimentali di un sistema televisivo via cavo in HF. (c)

12. Calzolari P., Paladin G.: Considerazioni sui cavi per diffusione televisiva. (c)

13. Canato L., Kordalis G., Scozzari G.: Apparecchiature di trasmissione per la distribuzione di segnali televisivi nel sistema HF. (c)

14. Bassani G., Marchetti G.F.: Matrice di commutazione e logica di comando nel sistema HF. (c)

15. Paladin G., Savino A.: Evoluzione di una rete di distribuzione TV in banda HF e sua integrazione con la rete di telecomunicazioni. (c)

16. Billia G., Ruggi d'Aragona F., Savino A.: Sistemi di diffusione di informazioni sonore e televisive. (d)

17. Del Prato P.: Possibili servizi ausiliari connessi alla distribuzione in un sistema CATV-HF e relativi apparati. (d)

(a) Atti XXI Congresso Internazionale per l'Elettronica - Roma '74
(b) Atti della Tavola Rotonda - 75^ Riunione Annuale AEI - Roma 1974
(c) Atti del XXII Congresso Internazionale per l'Elettronica - Roma 1975
(d) Atti XXIII Convegno Internazionale delle Comunicazioni - Genova 1975

APPENDIX

Technical parameters of a CATV-HF system

The main component parts of a complete CATV-HF transmission system can be scheduled
as follows:

- modulator
- amplifier
- cable (microcoaxial and symmetrical screened pairs)
- switching matrix
- frequency converter
- colour receiver

In order to have satisfactory quality performances by the overall system reference
has been made to the quality obtainable with broadcasting system, in the service
area of the main transmitter.

To realize this goal it is necessary to use transmission devices with very high
technical performances.

The following data were measured on the developed apparatus.

Modulator

Amplitude-frequency response [*]	$\pm 0,25$ dB (up to 4.8 MHz)
Group delay distorsion [*]	± 30 ns (up to 4.8 MHz)
Differential phase	3,5°
Differential gain	1,5%
K factor	2%
Chrominanceluminance gain inequality	2%
Chrominance luminance delay inequality	20 ns
Signal to noise ratio (weigted)	66 dB

Transmission line [**]

Amplitude-frequency response	$\pm 0,5$ dB
Group delay distorsion	± 30 ns
Single amplifier noise figure	3,5 dB

---

[*]   Measured with a test demodulator

[**]   Composed with 12 amplifiers and an overall lenght of 24 Kms microcoaxial
        cable

Video Switching matrix

(see pag. 4 of the text)

Overall transmission system***

Amplitude–frequency response	± 1 dB
Group delay distorsion	± 60 ns
Field time waveform distorsion	0,5 %
Line time luminance signal non linear distorsion	2%
K factor	3%
chrominance luminance gain inequality	6,5%
Chrominance luminance delay inequality	20 ns
Differential phase	2,5°
Differential gain	1%
Signal to crosstalk ratio	55 dB

Trunk line cable

The 0.7/2.9 mm microcoaxial cable used since many years in the Italian medium bit
rate digital network, is also suitable for transmission of TV signals in the main
and secondary network.

This cable, described in various papers   or in the Pirelli's catalogue
"Coaxial pair cables 0.7/2.9 mm for digital medium bit rate transmission", is manufactured at present with capacity up to 48 microcoaxial pairs. It has either lead
alloy or alluminium sheat and outer polyethylene jacket for direct burial or duct
laying. Steel tape armouring may be used for mechanical and electrical protection.
Microcoaxial pairs, at present under standardisation by C.C.I.T.T., have such cha
racteristics as to achieve very low levels of cross-talk between pairs.
It is possible to transmit over each pair a television signal allocated in the HF
band (2.5 to 9.5 MHz) up to the electronic devices; television signals can be transmitted in the same cables e.g.; with more PCM systems using different pairs.
Links using microcoaxial cables can have maximum lenght from the head end of approx.
20 km with a repeater spacing of 2 Km, or 30 km with a repeater spacing of 1 km.

User's distribution network

At the output of the switching matrix, the television signal is transmitted over
special symmetric pairs also together with telephone signals, wire broadcasting, data

*** Measured with a test demodulator substituting the television receiver and video—telephone signals, which are so simultaneously fed to the subscriber within a band of approax. 14 MHz. In the following Table the characteristics are given of an experimental cable having 10 screened pairs, 0,6 mm polyethylene insulated copper conductors, aluminium screen on the core and outer polyethylene jacket.

Cable with 0.6 mm screened pairs with PE or PVC sheat		
n° of pairs	10	1(subscriber terminal distribution
PE covering thickness mm	1.6	–
PVE covering thickness mm	–	0.7
Outer diameter mm	15	4.3
Weight per km kg	200	23
Nominal legth m	500	250 (coils)

- Capacitance at 800 Hz			max 61 nF/km	
- D.C. Voltage between conductor and screen (1 min)			600 V	
- D.C. Electrical Resistance at 20°C			max 62,1	
- Attenuation at 20°C MHz	2	5	10	15
dB	36	53	70	73
- Equal level F.E.X.T.attenuation (10 pairs – lenght 550 m) values measured from 1 MHz to 20 MHz more than			60 dB	
- N.E.X.T. attenuation (10 pairs – length 500 m) worst value measured from 1 MHz to 20 MHz more than			70 dB	

Ein in Italien erprobtes HF-Kabelfernsehsystem mit der Möglichkeit eines Rückkanals

Ein kürzlich in Italien entwickeltes und erprobtes HF-Kabelfernsehsystem wird beschrieben. Die wichtigsten Merkmale dieses Systems sind eine möglichst weit-
gehende Ausnutzung der vom bestehenden Fernsprechnetz angebotenen Betriebs-
einrichtungen und eine Studie über die Möglichkeit, die Kabelfernsehanlage zu-
sammen mit anderen Nachrichtenarten, wie z. B. Datenübertragung, Bildtelefon,
Drahthörfunk und selbstverständlich Fernsprechen, zu betreiben.

Das System ist sternförmig aufgebaut. Die Wahl des gewünschten Programms
erfolgt durch Betätigen einer Wähleinrichtung mit Fernsteuerung, die die Teil-
nehmer-Anschlußleitung an das gewählte Programm anschaltet.

Die von dem System angebotenen Möglichkeiten, Kanäle in Rückrichtung vor-
zusehen, werden besprochen.

The Combined Use of CATV- and Telephone-Networks for Purposes of Education and Consultion

Jan L. Bordewijk
Delft, Netherlands

1. Introduction

The purpose of this contribution is to elucidate certain philosophies with regard to the realization of low cost education and consultation services by a combined use of CATV- and telephone networks and to report on some initial achievements in this direction. These include the design and installation of a so-called double-star CATV-network {1} the development of an electronic blackboard or "teleboard {2} and first performance observations. Messages written on the "teleboard" can be either transmitted simultaneously with speech over an ordinary telephone circuit or as a standard television signal over a CATV-channel.

2. CATV a potential tresspasser

It is often stated, and generally accepted as a matter of course, that CATV-networks can be used for quite a number of applications other than the service of distributing broadcasting programs. Many endeavours, however, to design, introduce and develop such new services have suffered shipwreck in spite of optimistic expectations.
One may wonder whether these manifold reverses are due to a series of unfortunate coincidences or whether symptomatic phenomena are playing a role.

One such phenomenon might be that cable television is a highly controversial topic in the world of communications. Cable television by its very nature is a potential tresspasser of well-established demarcation lines. Such demarcation lines are found not only in the area of software production, as between broadcasters, press, publishers, educationalists and so forth, but also in the field of tele-communication between cable-engineers and radio-engineers, etc.

In an attempt to ward off a possible intruder, some telecommunication-authorities offer the opinion that the function of cable television should be restricted to the distribution of broadcasting programs.
Others, overestimating the present state of technology, glorify cable television networks as the new and final pathways of our future. Their slogan "wired city" and the almost canonized yell: "two-way" evoke in their turn for many people the spooks of "Big Brother", information pollution and so on.

Broadcasting organizations and artists' unions follow the development of cable television Argus-eyed. Newspaper agencies and publishers suspect that cable television may eventually carry part of their software production directly to the consumer.

Secondly people are sometimes too easily inclined to believe that concepts valid in one domain can be transposed into a related field with equal success.

A case in point is the manner in which successful transmission of moving pictures in television broadcasting has stimulated attempts at (too) early introduction of the transmission of moving pictures in dialogue communication systems in the form of the videophone.

In the light of such varied concerns, conflicts of interests and transposition dangers it seems logical, in studying alternative uses of CATV-networks from a technical point of view, to draw the whole terrain of (audio)visual (tele)communication into the discussion. One must not scrap in advance the possibility of adjustments to demarcation lines and combination possibilities.

3. Classification of (audio)visual (tele)communication

Audiovisual telecommunication systems can be divided into three main categories

- distribution systems
- consultation systems
- dialogue (c.q. polylogue) systems

In each of these categories we can discriminate between analogue brightness "visuals" indicated as "pictures" in this contribution and digital brightness "visuals" indicated as "graphics". Pictures as well as graphics can again be divided into dynamic (moving) or static (still).

By applying such a two-fold division we obtain four different types of "services" in each of the three categories mentioned above:

- moving pictures (normally with sound)
- still pictures (sound facultative)
- moving graphics (sound facultative)
- still graphics (normally without sound and alphanumerical)

In fig. 1 the 12 types of services arrived at in this way are tentatively arranged and nominated.

In the pure distribution category, moving pictures represent the well-known services of TV-broadcasting by air or by cable: TV programs offered according to a fixed time schedule. According to our definition, even dial- and pay-television as proposed for CATV-networks fall into this category so long as the timing of TV programs is fixed in the distribution centre.

In the consultation category, moving pictures refer to the film- or videolibrary (videothecque).

In the dialogue category under moving pictures we find the videophone.

It is indeed striking that following upon the success of television broadcasting interest turned immediately to a straightforward transposition into the consultation and dialogue categories.

It is now wellknown that in the absence of both long distance wideband transmission infra-structures and wideband switching infrastructures the videothecque and especially the videophone service will take many years to become established as a large scale public facility.

In all three categories attempts were undertaken to create low cost services. One way of reducing cost is to restrict oneself to still pictures; a second solution consists of restricting oneself to graphical "visuals".

A double saving in this respect is obtained by the services mentioned in the "lower righthand" quadrants of the three categories arranged in fig. 1.

Both reductions have also some important educational or, expressed in a more general way, information-transfer consequences apart from their cost-saving aspects.

Graphical illustrations as an abstraction of reality are well-known for their didactical value.

Dynamographic "visuals" as used by teachers in drawing and writing on blackboards, overhead-projectors etc. are often of still greater didactical value.

Most fortunately reduction of information is not equivalent with less effective information transfer as we can learn from the presence of information reducing stages in human perception systems {3}.

4. Cost savings in actual projects

The introduction of still-picture broadcasting as studied by N.H.K. in Japan{4} the use of the electronic blackboard by the Fryske Akademy in the Netherlands {5} and the well-known text-broadcasting by means of Ceefax and Oracle in the United Kingdom provide for cost reducing solutions in the distribution category. In both the N.H.K. still-picture broadcasting experiments and the text-broadcasting services in the U.K. savings are obtained by an improved adaptation between the "software problem" on the one hand and the available information capacity of the transmission channel on the other hand.

One could call this kind of adaptation: "generalized source coding". A second important cost saving factor is to be found in the reduced program production costs.

The Fryske Akademy uses a complete TV-channel to broadcast an audiosignal plus a writing signal destined for a hundred primary schools in Friesland. The information rate of the writing signal can be reduced with an adaptive differential chain coding {6} to approximately 200 bit/sec. The cost reduction in this case is of a different nature. It is achieved by savings in program-production (a simple audio studio is used) as well as by the avoidance of interface equipment at the receiver sites. The conversion into a standard 625-lines TV-signal can be centralized at the transmitter. Fig. 3 shows the scribosonic studio and classroomreception.

The cost reductions apply equally as well to broadcasting as to cable casting.

5. A low cost education service in CATV-networks

The foregoing considerations may guide us in designing low cost services in ČATV-networks.

CATV-networks exhibit qualities than cannot be obtained by means of broadcasting over the air.

The following guide lines may be of some help

1. - Choose cheap software sources
2. - Utilize the CATV-network as a distribution structure only
3. - Exploit the wide-band properties of CATV
4. - Exploit the inter-activity and routing potentials of the telephone network.

ad 1 Choosing cheap software sources means avoiding the use of complete television studio's id est: using still-picture, alpha-numeric information, teleboard and so forth.

Low cost per unit of time could be reached by continuous repetition of suitable programs as for example education courses, children's programs etc.

ad 2 Installation of reverse channels in a CATV-network is for the time being a costly affair.

ad 3 By allocating a full TV-channel for low information rate signals expensive interface equipment at the receiver site can be avoided.

ad 4 By using the telephone network for reverse information the use of an existing infrastructure is improved.

In those countries that possess sufficiently dense and well designed CATV-networks the transmission of still pictures and moving graphics as well as still graphics may well be economically feasible within a short period of time.

A proposal for a socio-technical pilot project, in the town of Zaltbommel in the Netherlands, is presently being seriously considered and is based on the following elements.

a. The presence in Zaltbommel of a so-called double-star CATV-network in which each home is connected to a "star-center" by means of two thin coaxial cables of the C-18 type (18 dB/km at 230 MHz) (fig. 2) {7}

Approximately 60 homes are connected in this way to one "star center". The "star-centers" are connected by a trunknetwork to the head-end of the CATV-network. By means of the second cable, each subscriber can be connected individually to an "education or consultation center" to form there together a kind of "tele-class" with the teacher in the tele-classer center. The "tele-class" network forms a closed-circuit to which only teacher and "pupils" have access. The investment cost of this kind of CATV-networks has turned out to be only 10% higher than that of the classical "branching-off" type networks. Due to a more favourable ratio between passive and active components the subscription rate stays approximately the same. As the second cable can also be used for a second packet of distribution type programs existing TV-receiver sets with a restricted provision of preset channel selection can continue in use. This more than compensates for an eventual slight increase in subscription rate. .

Cheap electronic "packet-switches" with a capacity for packets of 1 to 6 TV programs have been developed and can be inserted in the "star-centers". The electronic switches can be installed for each subscriber individually at his request. For the first experiments manual semi-permanent through-connection is considered.

b.   The development of a simple writing-tablet {6} (see fig. 4)

c.   The development of a multiplexer {8} that accomodates an audiosignal and a 200 bit/sec writing signal into the standard telephoneband of 300-3400 Hz. This development is part of an experimental scribophone project in which standard "dial-up" telephone-circuits are used to transmit the so-called "scribophone signals".

d.   The development of a convertor {6} that converts the "scribophone signals" into a standard 625-lines monochrome or colour TV-signal (fig. 5).

By providing the teacher and each pupil with a writing tablet and a multiplexer, by connecting each pupil by means of a telephone conference circuit with the teacher and the other pupils and by providing only the teacher with a conversion-unit whose output is connected to the closed-circuit part of the CATV-network, a tele-class is created in which each participant can speak and write to every other participant within the privacy conditions of a normal classroom environment.

It is felt that in such a system a balanced use is made of the properties of existing infra-structures.
The only costly element, the conversion unit, appears only once in the system!

Inter-activity so desirable in education and consultation situations is obtained by making use of the property of easy inter-activity inherent to telephone-networks, while the necessity of a conversion-unit per home is avoided by using the CATV-network.

The operation of the teleboard and conversion-unit is of such simplicity that it can be operated by the teacher from his home or from a simple audio-studio. The broadcasting experiences obtained by the Fryske Akademy {5} demonstrate this conclusively.

Literature reference

1. Krijger, L., Lichtenberg, B.C.: Multifunctional CATV-system. Cable television engineering, Aug. 1975

2. Kegel, A. et al.: The electronic blackboard. Tijdschrift voor het Nederlands Electronica- en Radiogenootschap, deel 38, No. 6, 1973

3. Marko, H., Giebel, H.: Recognition of handwritten characters with a system of homogeneous layers. NTZ Heft 9, 1970

4. Ando, H., Yamane, H.: Still picture broadcasting - a new informational and instructional broadcasting system. IEEE Transactions on broadcasting, Vol. BC-19, No. 3, Sept, 1973

5. Fryske Akademy: Het gebruik van het teleschoolbord in Friesland, Dec. 1975

6. Kegel, A., Bons, J.H.: On the digital processing and transmission of handwriting and sketching (Eurocon, Venezia, May 1977)

7. Bordewijk, J.L.: On the marriage of telephone and television. IEEE Transactions on communications, Vol. Com-23, No. 1, January 1975

8. Internal Reports, Delft University of Technology

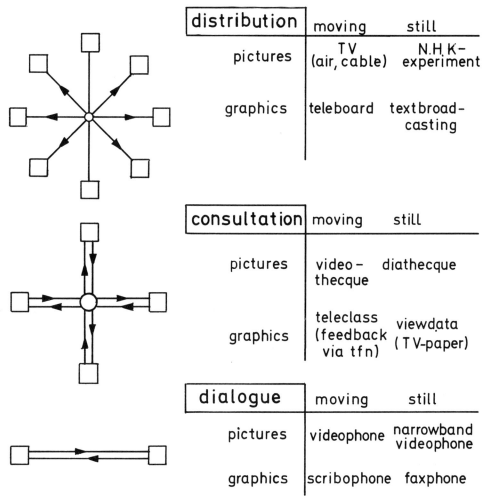

distribution	moving	still
pictures	TV (air, cable)	N.H.K- experiment
graphics	teleboard	textbroad- casting

consultation	moving	still
pictures	video- thecque	diathecque
graphics	teleclass (feedback via tfn)	viewdata (TV-paper)

dialogue	moving	still
pictures	videophone	narrowband videophone
graphics	scribophone	faxphone

fig.1 Survey of (audio) visual (tele) communication systems

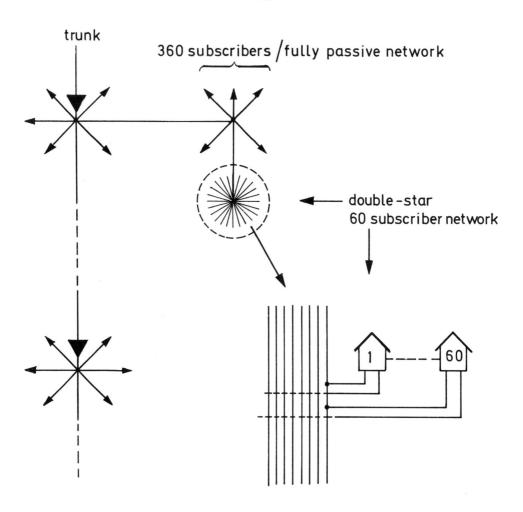

fig.2 VHF – double – star CATV – network

fig. 3. Scribophonic studio
and classroomreception

fig. 4. Electronic writing tablet

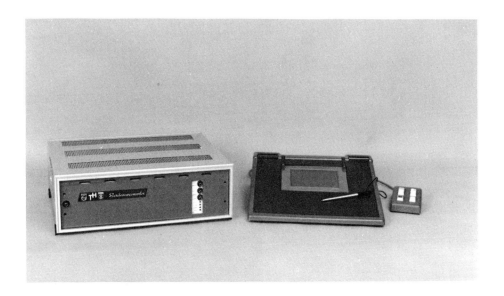

fig. 5 Convertor and alternative writing tablet

Der kombinierte Einsatz von Kabelfernseh- und Fernsprechsystemen beim beruflichen Informationsaustausch und im Bildungsbereich

Der Zweck dieses Beitrags ist es, verschiedene Konzepte für die Realisierung kostengünstiger Bildungs- und Dialogdienste durch die gemeinsame Nutzung von Kabelfernseh- und Fernsprechnetzen auszuleuchten und erste Erfolge in dieser Richtung vorzustellen. Dazu gehören die Konstruktion und Errichtung eines sogenannten Doppelstern-Kabelfernseh-Netzes, die Entwicklung einer elektronischen Wandtafel ("Teleboard") sowie erste Betriebsstudien. Schriftzüge auf dem "Teleboard" können entweder gleichzeitig mit dem gesprochenen Wort über herkömmliche Fernsprechleitungen oder als reguläres Fernsehsignal über einen Kabelfernsehkanal übertragen werden.

The Integrated Digital Telephone System –
a Possible Alternative to a Wide-band Network

Walter Neu
Bern, Switzerland

Abstract

A wide-band distribution network with backward channels is a combination
of a distribution network with a switching system. The latter resembles
a telephone switching system, but its cost-determing properties such as
traffic or geographical density of subscribers, are largely unknown. It
is therefore tempting to utilize the already existing telephone network
at least for the backward channels. In applications that do not call for
moving pictures it might even be feasible for the forward channel. Now,
it is true that even the transmission of still pictures over telephone
lines is unsatisfactory due to its slowness. On the other hand, the
emerging digital telephone systems could improve that situation considerably: firstly because a digital speech channel is inherently better
suited to picture transmission, secondly because in a digital system it
is relatively easy to form multiple channels, and thirdly because the
digital switching system provides almost unlimited flexibility in the
number and distribution of programmes and participants.

The need for telecommunication

Today we have mainly two telecommunication systems: telephony and broadcasting. (In fig.1 these have been denoted by A and E, respectively).

Now, it seems fairly obvious that in the future there will be a requirement for forms of telecommunication that are in between those two extremes.

In their comments on the KtK-report, the federal government of Germany
have rightly noted that the decisions on future telecommunication systems
will have a strong bearing on such topics as "diversity of opinion" or
"freedom of information", which are basic for a free democracy.

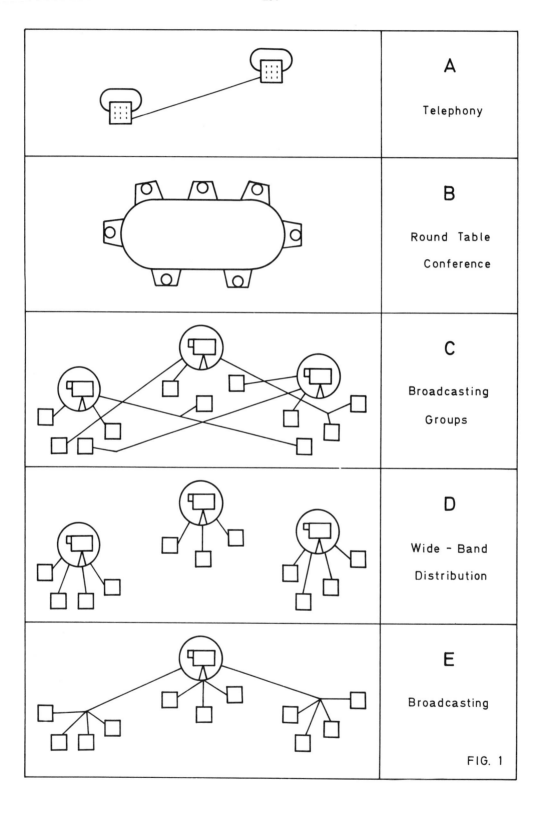

A

Telephony

B

Round Table Conference

C

Broadcasting Groups

D

Wide - Band Distribution

E

Broadcasting

FIG. 1

Even without entering into these political considerations it is easily
seen that there is a need for telecommunication between groups of people
that are too large for a telephone connection and yet too small for
broadcasting to be feasible. One example is that of continuing education.
Courses given at schools and universities are all right but they alone
will hardly solve the whole problem. Some help by telecommunication
facilities seems to be essential.

The problems

Wide-band distribution networks with backward channels represent one
line of approach. They are particularly attractive in those cases in
which the group of people to be served by a programme live all in the
same local community (case D in fig. 1).

There are many applications, however, with groups of participants distributed over the country. For continuing education, for example, there
may be just one or two specialists in each community who are interested
in a particular course, as sketched under C in fig. 1.

Whereas in case D the handling of backward channels would be fairly
simple, in case C it is likely to become a task of similar magnitude to
that of a telephone switching system. The question arises, therefore,
if it would not be better to use the already existing telephone network for that purpose.

Also the transmission of forward (broadband) channels creates problems,
if case C is presumed, with its multitude of programme sources, each
serving a restricted number of receivers with a broad geographical distribution.

The problem becomes even more complicated if another important application is considered, which we have called "round table conference"
(case B in fig. 1). Here, each member of the group should have the same
opportunity to speak and to send text or pictures. As a specific example,
let's imagine a CCITT conference with an attendance of the order of
100 including one or a few delegates from each country. This is a situation which is even more remote from broadcasting (E or D); nor is it
properly manageable with the present telephone system.

In view of these difficulties, it really seems that only a new type of
switching network could adequately meet the requirements in the long run.

For economical reasons, however, it is doubtful if a broadband switching network could be successfully introduced in the not too distant future.

It might require investments per subscriber of the order of 10 times those of the telephone network, according to estimates in the KtK report. The high cost is of course mainly caused by the need for transmission of moving pictures. The next question, therefore, is whether moving pictures are essential or still pictures would suffice. The KtK report concludes that for most applications still pictures might be adequate.

What sort of network is required, then, if only still pictures along with speech and data have to be transmitted? It has been mentioned that the telephone network might do that. It can easily be calculated, however, that the transmission of a single picture would take between one and several minutes. This, we think, is too slow.

Also, the present telephone systems are not capable of providing connections of types B or C in fig.1, except to a very limited extent.

The digital telephone network

The situation changes if we introduce digital telephone systems which make use of pulse code modulation. They switch digital channels of 64 kbit/s which can provide much more efficient picture transmission. It is even possible to use multiple channels of, for example, 128 kbit/s. This brings the time required for one picture down to something like 5 seconds. Moreover, digital switches allow connections of types A,B, C,D,E (fig. 1) to be made almost without restrictions.

The question is now whether these promising possibilities justify the introduction of integrated digital networks or not. An answer to this question would normally call for some pseudo-scientific procedure like market research for a product that does not yet exist. Fortunately, however, it does not seem to be necessary in this case, because the introduction of a digital communication network appears to be economically justified already by telephony alone. At least, this is the conclusion reached by the Swiss PTT and the three firms Hasler, Siemens-Albis, and Standard Telephon & Radio after some 5 years of joint research and development. Their system "IFS" uses digital transmission and switching throughout with the exception of the local network (concentrator and subscriber line).(1).

According to recent estimates, assuming full introduction, digital switching exchanges should be approximately 35% cheaper than present electromechanical exchanges. Even during the first introduction stages where costly adaptions to existing systems are needed, they promise to be competitive. New programme controlled electromechanical exchanges, on the other hand, would in all probability be more expensive than our present systems.

This shows that in order to prepare the ground for a comprehensive future telecommunication network, it is not necessary to go for large investments with uncertain return. All that is required at first is to select the proper direction in the development of the ordinary telephone system.

The additional effort required for the new services and applications lies then in the comparatively modest task of developing a digital concentrator and subscriber line, some more software, and of course suitable terminals for the subscribers.

A working party formed by the smaller Swiss telecommunication firms (Autophon, Gfeller, Zellweger) and the PTT are now developing an experimental network model including digital concentrators. These are not yet designed to be connected to the IFS system but are self-contained units used for demonstration and evaluation of the new services, together with a novel type of "display telephone" as terminal.(2).

The IFS switching unit

It may be appropriate now to explain why the digital switching system readily allows connections of types B and C to be made, while our present telephone systems do not.

The basic switching unit is in effect what computer people call a "random access memory". It can interconnect 28 PCM-lines each carrying 30 speech channels (2.048 Mbit/s). One 8-bit word of each channel is temporarily stored in the memory, from which it can be read out and sent to another channel of any of the 28 multiplex lines. Since only one channel is handled at a time and the same 8-bit word can be read any number of times, each incoming channel can be connected to any number out outgoing channels, thus forming a "broadcast" connection of type C, D, or E.

Moreover, the two directions of transmission can be switched independently, so that loops for round-table conferences can be formed (case B).

These large non-blocking switching units have another advantage, in that they allow the design of switching networks with practically unlimited traffic capacity per subscriber. This means that many or all subscribers could be connected simultaneously; a situation that might well occur in the broadcast case (C to E).

Conclusions

It has been noted that, while a wide-band switching network might be a desirable future telecommunication medium, it does not seem to be economically feasible for some time to come.

On the other hand, excepting the transmission of moving pictures, all foreseeable requirements could presumably be met by an augumented digital telephone network. This opens the prospect of worldwide telecommunication between groups of people of any size and distribution. Services would include : good quality speech, still pictures, facsimile, text and general data.

Bibliography

1. Burger, Peter A. : System IFS - The Swiss Approach to Digital Communications. Communications Society, IEEE, November 1976, Vol. 14, No.6'

2. Kündig, Albert : Ein experimentelles Endgerät für die Sprach-und Bildübertragung in PCM-Netzen. Technische Mitteilungen PTT, Bern, 1976, No. 11 and 12.'

Das integrierte digitale Fernsprechsystem – eine mögliche Alternative zum Breitbandnetz

Ein Breitbandverteilnetz mit Kanälen für die Rückrichtung stellt eine Kombination aus einem Verteilnetz und einem Vermittlungssystem dar. Das letztere ähnelt einem Fernsprechvermittlungssystem, dessen kostenbestimmende Faktoren, wie die Verkehrsdichte bzw. die räumliche Dichte der Teilnehmer, aber weitgehend unbekannt sind. Dabei wird häufig erwogen, das bereits vorhandene Fernsprechnetz zumindest für die Rückrichtung auszunutzen. Bei Anwendungsfällen, die keine Bewegt-Bilder erfordern, könnte es sogar für die Verteilrichtung verwendet werden. Allerdings ist selbst bei der Übertragung von stillstehenden Bildern auf den üblichen Telefonleitungen auf Grund der geringen Übertragungsgeschwindigkeit kein zufriedenstellendes Ergebnis zu erzielen. Andererseits aber könnte durch das Aufkommen digitaler Fernsprechsysteme die Lage entscheidend verbessert werden: erstens, weil ein digitaler Sprachkanal für die Bildübertragung besser geeignet ist, zweitens, weil die gemeinsame Nutzung mehrerer Zeitmultiplex-kanäle in einem digitalen System relativ einfach ist, und drittens, weil bei einem digitalen Vermittlungssystem die Möglichkeiten bezüglich der Anzahl und Verteilung von Programmen auf die Teilnehmer nahezu unbegrenzt sind.

The Commission on the Future of Telecommunications in the Federal Republic of Germany (KtK) – its Task and its Recommendations Regarding Wide-band Communications

Wolfgang A. Kaiser
Stuttgart, Federal Republic of Germany

Abstract

The German Commission on the Future of Telecommunications has concluded its assignment early in 1976 and submitted its report to the German Government. In this report not only the demand and the needs for new telecommunication services are treated but also the possibilities of their technical realization and their related costs. The paper gives a short review of the results and recommendations of the Commission particularly emphasizing broadband distribution networks. The Commission could not recognize a very pronounced and pressing demand for more television programs and therefore refrained from recommending a nation-wide broadband cable distribution network. On the other hand the many new forms of telecommunication possible in broadband distribution networks couldn't be disregarded. The Commission therefore recommended implementing pilot projects to test these new possibilities, their technical realization and alternative forms of organization. Several such projects are now in the planning stage.

1. Introduction

In 1974 the Government of the Federal Republic of Germany appointed the "Kommission für die Entwicklung des technischen Kommunikations-systems (KtK)" to submit proposals for a socially desirable and eco-nomically reasonable telecommunication system of the future. The Com-mission consisted of members belonging to various groups such as poli-tical parties, federal and local authorities, trade-unions, press, broadcasting, trade, industry and science. It organized its work by forming four working parties which dealt with the needs and demands for telecommunication, the technologies and costs, the organizational problems and the financing of the future telecommunication systems.

After two years of work the Commission completed its assignment. The results are laid down in the "Telecommunications Report" and eight annex volumes (1,2). They cover the general conclusions and recommendations and give detailed information on the four areas of interest which have been considered:

- Existing forms of telecommunication

- New forms of telecommunication with existing networks

- Forms of telecommunication with broadband distribution networks

- Forms of telecommunication with switched broadband networks.

In July 1976 the Federal Government published its answer (3) to the recommendations of the Commission. It declared to take up and implement most of the proposals given and has already taken steps to achieve these innovations.

2. Telecommunication in existing networks

The existing telecommunication services, namely telephone, telex, data communication, mobile service, radio and TV have reached a high standard in Germany but leave room for further development. Telephony will continue to be the most important form of telecommunication. However, the telephone density is still comparatively low. Thus it is recommended that the telephone network is extended with the aim of providing full coverage of all households in the near future. The respective efforts made by the German Bundespost in the meantime have already been very successful.

The growth rate of data communication is considerable. Therefore it is recommended that the datel services are further developed and that the integrated data network now introduced be completed soon.

Remarkable innovation effects may be achieved by implementing new forms of telecommunication in the networks already existing. From the long list of possibilities (Fig. 1) studied by the Commission special forms of text and still picture communication such as communication typewriting and telecopying services for office subscribers as well as the extended use of TV-sets for the private subscribers seem to be of greater interest. The German PTT has taken steps to realize many of the recommended new forms of telecommunication. The text communication forms called VIDEOTEXT and VIEW DATA (Bildschirmtext), which are using the TV screen for the reproduction of the text picture, are planned to be demonstrated to the general public at the Broadcasting Exhibition in Berlin in August 1977. VIDEOTEXT, similar to the British Teletext, is the transmission of text by using blank lines of the TV-signal, whereas VIEW DATA (Bildschirmtext) makes use of the telephone network for the text transmission. Thus it does not only allow the distribution, but also the retrieval of text messages.

	FORM OF TELECOMMUNICATION	FLOW OF INFORMATION		NETWORK	OUTPUT ON
TEXT	COMMUNICATION TYPEWRITER		DIALOG	DATA- OR TELEPHONE-	PAPER
	VIDEO TEXT		DISTRIBUTION	TV- OR CATV-	SCREEN
	VIEW DATA (BILDSCHIRMTEXT)		RETRIEVAL	TELEPHONE -	SCREEN
STILL PICTURE	FACSIMILE		SUBSCRIB.TO SUBSCRIBER	TELEPHONE- OR DATA-	PAPER
	FACSIMILE NEWSPAPER		DISTRIBUTION	TELEPHONE- OR CATV-	PAPER
	VIDEO - STILL PICTURE		RETRIEVAL	CATV-	SCREEN
	SLOW-SCAN TV		RETRIEVAL	TELEPHONE -	SCREEN
TELE-METERING	ELECTRONIC MAIL		SUBSCRIB.TO SUBSCRIBER	TELEPHONE- OR DATA -	PAPER
	TELEPHONE CONFERENCE		CONFERENCE	MANUALLY SWITCHED TELEPHONE CONNECT.	TELEPHONE RECEIVER
	REMOTE CONTROL		CENTER TO SUBSCRIB.	TELEPHONE-, DATA- OR CATV-	REMOTE CONTROL EQUIPMENT
	REMOTE METERING		COLLECTION		
	● CENTER ○ SUBSCRIBER				

Fig. 1. New forms of telecommunication in existing networks

3. Forms of telecommunication with broadband distribution networks

In Germany many households are already connected to master or community antenna TV-systems. These systems might, if standardized, be extended and improved to form broadband cable distribution systems, which gradually might grow to become large networks covering wide areas. Such systems would allow the realization of Cable television which is defined as the distribution of additional TV-programs not receivable by normal antennas. At the moment it is not yet allowed for regulatory reasons. Establishing broadband distribution networks the problems encountered are not so much of a technical but rather of a financial, political, social and legal nature.

Since the large capacity in such systems - the Commission recommended
30 channels - is not completely used by TV-programs additional services may be offered. Fig. 2 lists some of the possible forms of telecommunication. The subscriber set necessary to receive these additional programs consists of a TV set adapted to the different applications by (normally built-in) special attachments.

FORM OF TELECOMMUNICATION		TYPICAL PROGRAMS	NECESSARY SUBSCRIBER SET
DISTRIBUTION	TELEVISION	NORMAL TV-PROGRAMS FOREIGN PROGRAMS PROGRAMS RECEIVED VIA SATELLITES TIME-SHIFTED PROGRAMS LOCAL PROGRAMS PROGRAMS FOR SPECIAL INTERESTS	TELEVISION SET CAPABLE TO RECEIVE THE REQUIRED NUMBER OF CHANNELS
	AUDIO	NORMAL PROGRAMS LOCAL PROGRAMS	RECEIVER
LIMITED DISTRIBUTION	TELEVISION	PAY-TV (SUBSCRIPTION)	TV-SET WITH ATTACHMENT A
		COIN-TELEVISION	TV-SET WITH ATTACHMENT B
	VIDEOTEXT	TEXT NEWS AND MESSAGES SUBTITLES	TV-SET WITH ATTACHMENT C
	VIDEO STILL PICTURE	CYCLIC SERIES OF STILL FRAME PICTURES	TV SET WITH ATTACHMENT D (INCL. PICTURE STORE)
	SLOW-SCAN TV	RETRIEVAL OF STILL PICTURES VIA THE TELEPHONE NETWORK	
	FACSIMILE NEWSPAPER	NEWS AND MESSAGES TO BE RECORDED ON HARD COPY	RECORDING UNIT

Fig. 2. Forms of telecommunication in one-way
broadband distribution networks

A tree-type network structure is the most economic way of creating a
broadband distribution system. Fig. 3 shows the cascading scheme standardized up till now. Coaxial cables transmit the signals in four hier-
archical levels (A, B, C and D) from the head-end or center of the
system to the interface points to which the private coaxial subscriber
lines are connected. Amplifiers are installed in the A- and B-level
only.

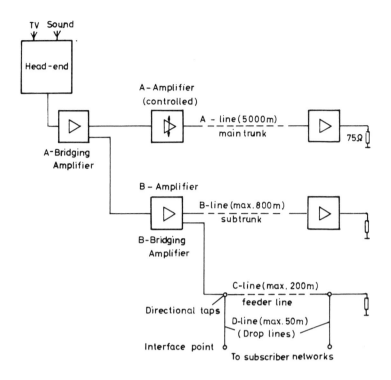

Fig. 3. Cascading scheme in standardized German
community antenna systems

Applying the normal channel spacing of 7 MHz a total of 30 TV channels
may be provided by using Band I (47 - 68 MHz), Band III (174 - 230 MHz)
and the special Midband 104 - 174 MHz and Superband 230 - 293 MHz. The
TV sets used today can only receive Band I, Band III and the UHF-Bands
IV/V (470 - 790 MHz) and often enough do not possess sufficient selection to allow the simultaneous utilization of every channel in Band I
and Band III. To be capable of receiving all 30 TV channels a special
cable television set has to be applied or some of the special channels
have to be converted to the UHF-Bands. This may be done at the last
B-amplifier or at the TV set.

Using a planning model (4) the Commission found that the total investment required to create local broadband distribution networks would
amount to approx. 9 000 million DM, if all towns with more than 20 000
inhabitants were to be served, and approx. 23 000 million DM would be
necessary for full national coverage. Fig. 4 gives the split-up of
these amounts showing that almost 75 % of the expenditure have to be
invested into the branches near the subscriber (C, D and S). The

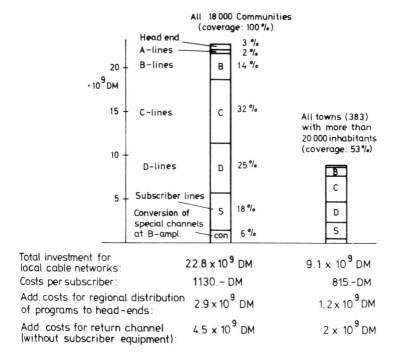

Total investment for local cable networks:	22.8×10^9 DM	9.1×10^9 DM
Costs per subscriber:	1130.– DM	815.–DM
Add. costs for regional distribution of programs to head-ends:	2.9×10^9 DM	1.2×10^9 DM
Add. costs for return channel (without subscriber equipment):	4.5×10^9 DM	2×10^9 DM

Fig. 4. Estimated investment necessary for the
installation of local broadband distribution
networks in the Federal Republic of Germany

investment per subscriber amounts to 650 ... 1 650 DM depending on
the size of the community he is living in. The regional and national distribution of programs to the head-ends of the local systems
is achieved at comparatively low additional costs, for instance by
utilizing the already existing 60 MHz coaxial cable systems or new
12 GHz microwave systems. Also, the use of a television satellite
seems to be both technically and financially feasible for this purpose.
The satellite, however, is no substitute for a local distribution
system because of the limited number of channels, the cost of private
receiving antenna systems and the inability to create an upstream
channel.

Upstream channels for two-way transmission enable the subscriber to
participate actively in the communication process. Fig. 5 gives an
over-all view of a local broadband system which uses forward and return
channels. For the return direction the frequency region below 47 MHz,
e.g. 5 - 30 MHz, may be used.

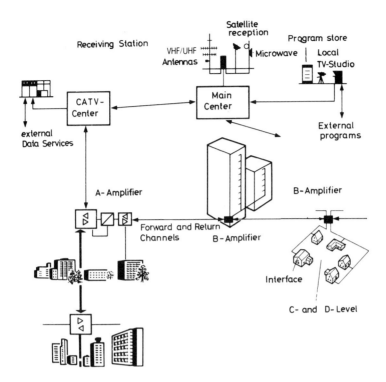

Fig. 5. Local broadband cable system with
return channels

Broadband cable communication systems open the way to many additional
forms of telecommunication (Fig. 6). The additional investment costs
for the network, for instance the two-way amplifiers, are comparatively
low as shown in Fig. 4. On the contrary the additional subscriber equipment is, even for simple applications, quite costly. Fig. 7 demonstrates
some of these possibilities. The frequency regions for the forward and
return direction are separated by a high pass / low pass filter. The
return channel uses the address code multiplex principle.

Adding an exchange to the center of the local broadband cable communication system allows the additional utilization of narrow band and
broadband channels for individual communication. Thus, elements of a
service integrated network could be established and tested. For instance, an acceptance test with VIDEO TELEPHONE might be started. Such
individual connections would mainly be restricted to the local network.
But the use of broadband connections between the head-ends would enable
regional and national transmission.

FORM OF TELECOMMUNICATION		TYPICAL SERVICES	NECESSARY SUBSCRIBER SET
RETRIEVAL	VIEW DATA	TEXT MESSAGES INFORMATION SERVICES LIBRARY ACCESS ORDER SERVICE	TV-SET WITH ATTACHMENT E
	RETRIEVAL OF STILL PICTURES	PICTURE .INFORMATION SERVICE, SLIDES	TV-SET WITH ATTACHMENT F
	RETRIEVAL OF MOVING PICTURES	BRIEF MOVIES FOR INFORMATION AND EDUCATION	TV-SET WITH ATTACHMENT G
COLLECTION	DATA COLLECTION BY CENTER	EMERGENCY CALLS, ALARMS, VOTING, OPINION POLLING PAY-TV-CONTROL	DATA SET FOR RETURN CHANNEL
	REMOTE METERING	SUPERVISION, UTILITY METERING	
DIALOG WITH NETWORK CENTER	VIEW DATA-DIALOG	INTERACTIVE EDUCATION AND TRAINING RESERVATION, ACCOUNTING	TV-SET WITH ATTACHMENTS E,F OR G
	DIALOG USING STILL OR MOVING PICTURES	REMOTE BUYING SOCIAL SERVICES MEDICAL CARE DIALOG WITH INFORMATION BANKS COMPUTER GAMES	

Fig. 6. Additional forms of telecommunication in
broadband cable networks with return channels

There are almost no bounds for the fantasy regarding broadband cable
systems with return channels, but for economic reasons the technical
possibilities are limited and the customer acceptance is still doubtful.

The Commission therefore recommended implementing pilot projects to
test alternative forms of telecommunication, their technical realiza-
tions and alternative forms of organization with the restriction that
the responsibility for the network should be separated from the one
for the programs. It suggested that the arrangements for experiments
within the framework of the pilot projects be as varied as possible
so that the acceptance and attractiveness of the use of broadband
distribution networks can be tested. As fas as broadcast programs are
concerned a decision on the part of the legally responsible Federal
States is indispensible. Several pilot projects are now in the planning
stage.

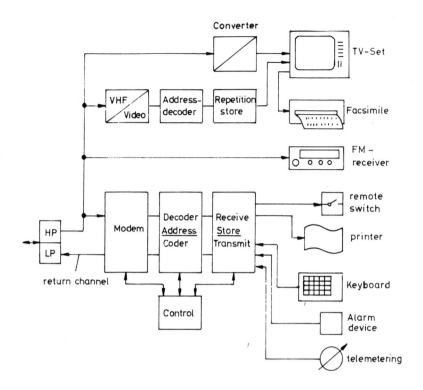

Fig. 7. Subscriber equipment for two-way broadband
cable systems

4. Switched broadband networks

People who are engaged in planning future telecommunication services
often hear the wish, that in a telecommunication system of the future
the subscriber should be able to communicate not only acoustically but
also by vision. But contrary to cable television where a distribution
network is sufficient, the bidirectional transmission of moving pictures between arbitrary subscribers, as it occurs in the video telephone service, the video conference or the transmission of individual
films on request, requires a public switched broadband network, for
instance in a star-type structure, to establish the many individual
broadband connections.

The typical difference between a network structure suitable for a
distribution network and one for a switched network is shown in Fig. 8.
Obviously a switched network needs considerably more investment. A
comparison of the two structures indicates also, that a broadband

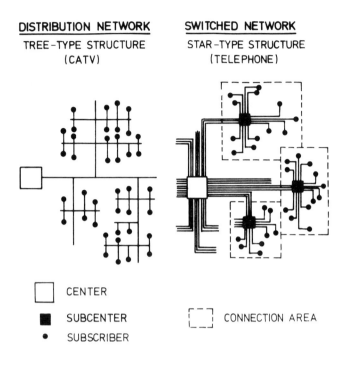

DISTRIBUTION NETWORK
TREE-TYPE STRUCTURE
(CATV)

SWITCHED NETWORK
STAR-TYPE STRUCTURE
(TELEPHONE)

☐ CENTER

■ SUBCENTER

• SUBSCRIBER

⌐ ¬ CONNECTION AREA
L _ ⌐

Fig. 8. Structures of networks

network with tree-type structure (distribution system) is not a preliminary stage for a star-type, switched network.

A switched broadband network can only to a small extent take advantage of the existing installation of the telephone network. Up to now such broadband networks have nowhere been tested to a great extent. Before their realization intensive research and development work has to be carried out to solve all the arising technical problems. But also the needs expressed by the subscribers have to be analyzed and related to the costs which they are willing to pay. Especially there is a large gap between the general desires and possibilities for future telecommunication on the one side and the economic power needed to realize it on the other.

In estimating the necessary costs four network models (A, B, C and D) have been considered by the Commission which reach from conventional wideband analog transmission to digital transmission and the use of optical communication on glass fibers. Depending on the selected

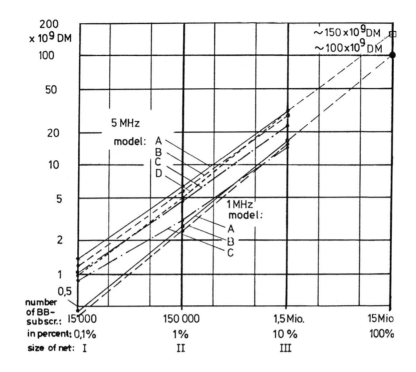

Fig. 9. Estimated investment costs for a nation-wide
switched broadband network for the transmission
of black/white video telephone signals with
1 resp. 5 MHz bandwidth

network model and the demands for quality of the transmitted pictures
(black and white TV with a resolution corresponding to 1 or 5 MHz bandwidth or color-TV with 5 MHz bandwidth) the report shows investment
costs between 100 and 300 $\cdot$ 10^9 DM for a nation-wide service for
15 million subscribers. Fig. 9 shows the estimated investment costs
for different numbers of subscribers. A broadband network with switched
connections is needed for the transmission of moving pictures only.
Of course it could additionally transmit all narrowband forms of telecommunication (voice, text, data, still pictures) which however can
be transmitted on the already existing networks too. In view of the
many open questions and the high financial investment needed the
Commission came to the conclusion that the introduction of a general
public switched broadband network is not yet justified.

5. Conclusions

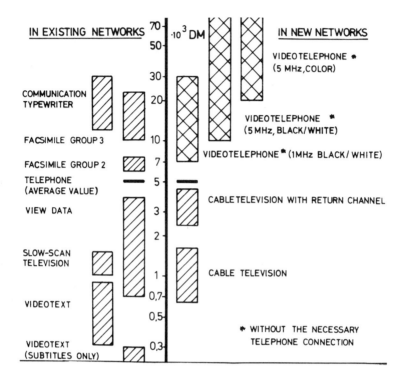

Fig. 10. Comparison of the investment per subscriber
for different forms of telecommunication

Considering the future possibilities of telecommunication three levels
of innovation may be distinguished:

- In addition to the conventional services the existing public networks
  may be utilized for new forms of telecommunication, especially for
  the transmission of text and still pictures. Such services might soon
  be introduced. The widespread use of these new forms is therefore
  less depending on the networks but mainly on the attractiveness of
  the service and especially the price of the subscriber set. The left
  hand side of Fig. 10 indicates typical investment costs per subscriber (incl. subscriber set). The lower values are valid for large
  numbers of subscribers and simple equipment. Since the average value
  of investment for a new telephone subscriber amounts to approx.
  5 000 DM, it may be assumed that the forms listed in the lower left
  part are of more interest to the private subscriber while the ones

in the upper left part are suitable for a business or office subscriber.

- For the transmission of moving pictures (e.g. TV) via cable new
  broadband networks are needed. The necessary investment for cable
  television and other forms of telecommunication in broadband distribution networks may be carried by the subscribers if the services
  offered are attractive enough. Since there are still many open questions pilot projects will be used as a test basis before final decisions are made.

- A switched broadband network needs considerable investment and may
  therefore be realized only in the remote future and after important
  technological progress (e.g. in optical communication) has been made.

References

1. Telecommunications Report (in English) and eight annex volumes
   (in German). Verlag Dr. H. Heger, Bonn, 1976

2. Kaiser, W.: Zukuenftige Telekommunikation in der Bundesrepublik
   Deutschland - Ergebnisse der KtK-Beratungen. Nachrichtentechn. Z.
   29 (1976), p. 190 - 210

3. Vorstellungen der Bundesregierung zum weiteren Ausbau des technischen Kommunikationssystems. Edited by Federal Ministry of Posts
   and Telecommunications, Bonn, 1976

4. Kabelfernsehen. Annex Volume 5 to the Telecommunications Report.
   Verlag Dr. H. Heger, Bonn, 1976

Die Kommission für den Ausbau des technischen Kommunikationssytems in der Bundesrepublik Deutschland (KtK) – Ihre Aufgaben und Empfehlungen im Hinblick auf den Einsatz von Breitbandsystemen

Die deutsche Kommission über die Zukunft der Telekommunikation hat ihre Untersuchungen Anfang 1976 abgeschlossen und den Bericht der Bundesregierung vorgelegt. In diesem Bericht wird nicht nur der Bedarf an neuen Telekommunikationsformen angesprochen, sondern auch die Möglichkeiten ihrer technischen Verwirklichung und die dabei entstehenden Kosten behandelt. Der vorliegende Beitrag gibt eine kurze Übersicht über die Ergebnisse und Empfehlungen der Kommission mit dem Schwerpunkt Breitbandverteilnetze. Eine besonders ausgeprägte und dringende Nachfrage nach mehr Fernsehprogrammen konnte die Kommission zwar nicht erkennen und sah daher von der Empfehlung, ein bundesweites Breitbandkabelnetz einzurichten, ab. Andererseits konnte man aber die vielen neuen Kommunikationsmöglichkeiten in Breitbandverteilnetzen nicht übergehen.

Daher empfahl die Kommission die Errichtung von Pilotprojekten, in denen diese neuen Möglichkeiten sowie deren technische Realisierung und alternative Organisationsformen erprobt werden können. Mehrere solcher Projekte befinden sich gegenwärtig im Planungsstadium.

Cable Television Pilot Projects in the Federal Republic of Germany

Jürgen Kanzow
Bonn, Federal Republic of Germany

A Opening remarks

I should like to try in the following to describe briefly the
present-day situation of cable television in the Federal Republic
of Germany from the point of view of the Deutsche Bundespost, and
the part played here by cable television pilot projects.

Professor Kaiser has just spoken to you about the findings and
recommendations of the KtK on the subject of broadband communication; he also mentioned the cable television pilot projects. After
lunch, State Secretary Professor Schreckenberger will talk to you
about the legal and political aspects connected with these projects.
Regarding its place on the agenda, therefore, my lecture falls in
the middle, which is certainly not a bad place for telecommunication!

B Current and planned cable television projects

0 Aim of the pilot projects

Professor Kaiser spoke in his lecture of the aims on which the recommendations of the KtK regarding cable television pilot projects
are based. For the Deutsche Bundespost, there are two questions of
particular importance. The first concerns the interest of the public
in cable television, and the second, the best possible structuring
of the technical system. Answering the first question is difficult,
because only in a few cases would cable television in the Federal
Republic of Germany also contribute to improving the reception of
existing television programmes. More than 95 % of the Federal

Republic of Germany is today provided with the three television
channels, and the remaining areas where reception is poor will
hardly be an immediate favourite for cable television facilities,
either. In other words, cable television cannot rely on the existence of a wide, basic demand for television. The attraction of
cable television derives primarily from the additional choice of
programmes which it alone can offer. However, this choice of programmes is still to a large extent unknown, and a factor about which
there is even more uncertainty is the interest of the public in such
a choice of programmes. We cannot predict this interest; moreover,
it would be of little use to ask the public without having given it
a demonstration beforehand of the subject of our inquiries. To make
possible such a demonstration is the chief task of the pilot
projects, and this is also the common interest of all those concerned with cable television. The second question, the best possible
structuring of the technical system, is a question above all for
the Deutsche Bundespost alone, as it is responsible for construction
of the cable television system, no matter whether it will also set
up and operate the network in the future.

1. Distribution networks

The Deutsche Bundespost has to date set up three so-called cable
television trial networks. These are in Nuremberg, Hamburg and
Düsseldorf. The facilities serve to resupply urban areas which are
shadowed by multi-storey buildings, where the reception of television programmes over the air is hindered or impossible. These
facilities do not transmit any programmes in addition to those
normally received over the air. Strictly speaking they are not,
therefore, cable television facilities, but are facilities used
to test the technical requirements which the establishment of a
nationwide cable television network will have to meet. At present,
the networks are operated purely as distribution facilities, even
if it is technically possible to provide return channels below
40 MHz. At the stage of complete development, these facilities
could supply an area of approximately 75 square kilometres, but
complete development has not been achieved in any of the three
trial projects. The situation is as follows.

	Nuremberg	Hamburg	Düsseldorf
supplied area in m^2	1,200,000	740,000	84,000
households within the supply area which may be connected	approx. 6,000	approx. 4,000	approx. 320
households at present connected (position as in March 1977)	2,470	1,560	281

1.2. The cable television network model of the Deutsche Bundespost

The structure of the trial facilities in Nuremberg, Hamburg and Düsseldorf is based on a cable television network model, very similar in its network topography to the topography of the public telephone network, so as, among other things, to be able to benefit from the already available infrastructure of the latter. In the cable television network model, a head station can supply an area with a radius of 5 kilometres from the head station. This corresponds approximately to the supply area of a local exchange in the telephone network. Within the cable television supply area the network is divided into four levels, of which the first two levels A and B are equipped with broadband amplifiers while the other two levels C and D are operated without. The D level ends at the so-called transfer point, situated on the border of the grounds of the house to be supplied or in the cellar of a house or block of flats. Further cabling within the grounds, within the house or block of flats, is the responsibility of the owner of the grounds.

In determining the present basic telecommunication values of the cable television network model we started from the assumption that the supply area of a head station would be regarded as it were as the basic element of a nationwide cable television network. A regional and supraregional cable television trunk network is set above these basic elements, shaping them to a nationwide, even cable television network and taking over the transportation of regional and supraregional cable television programmes. In addition, plans have been made for the head stations of the local cable television basic elements to receive, prepare and distribute in the connected network television programmes broadcast over the air. The amplifiers of the head stations and the broadband distribution amplifiers have been conceived in such a way that radio transmissions which can be received at the head station with the stipulated minimum received field strength, can be transferred at the transfer point with excellent quality.

Furthermore, the network model has been designed on the assumption that 12 distribution channels will be made available. As the frequency band available reaches approximately 300 MHz, it is presumably technically possible to increase the number of channels to 30, as recommended by the KtK. So far, only the form of reconversion of the cable television channels into broadcast bands remains to be decided, if we disregard for the moment the development of special cable television home receivers.

The experience gained in Nuremberg, Hamburg and Düsseldorf shows that this network model is possible and that the transmission qualities of the facilities are excellent. At the moment we are examining whether modification of some of the basic values is possible in order to attain even more economical solutions.

In conclusion, we may say that there are scarcely any problems from the technical point of view regarding the establishment of cable television networks which are used exclusively to distribute television programmes. Here, the pilot projects envisaged can only provide additional information if more than 12 distribution channels are occupied, and especially if the region of 30 channels, suggested by the KtK, is approached.

2. Two-way cable television

The interest of the Deutsche Bundespost is therefore concentrated on the return channel and the services which can thereby be implemented.

By return channel I mean in this context quite generally every sort of telecommunications path along which the transmission of signals from the subscriber to the head station, i.e. in the opposite direction to distribution, is possible. The type of signals to be transmitted determine whether the return channel must be narrowband or broadband. The channel must be available to the individual subscriber for shorter or longer periods of time, depending on the

nature of his participation. You will already realise that the
return channel cannot be dealt with generally, but that it can only
be defined in close connection with the return channel services
which may be implemented. For this reason, I should like to say
more about those questions which are connected for us from the network point of view with the individual groups of return channel
services.

2.1. Return channel services

First of all it is necessary to find a grouping for return channel
services that makes possible an evaluation of the technical requirements of the network. In view of the short time at my disposal,
I can only give you a very rough outline.

Many return channel services do not have identical technical requirements as regards the two directions. In the simplest case, the subscriber only has the possibility of saying "yes" or "no" to the
television programme showing. The most difficult case is certainly
presented by the individual transmission of moving pictures to the
subscriber and from the subscriber back to the cable television
centre. Between these two extremes lies the whole range of return
channel services under discussion in cable television networks.

Let us remain for the moment with the requirements made of the return channel in respect of performance. Four different types of
requirements may, I believe, be distinguished here:

- simple yes/no from all subscribers, simultaneously (vote)

- brief messages to the centre, individually, staggered (e.g.
  retrieval of information, pay TV)

- still frame pictures and longer texts to the centre, individually,
  staggered, (e.g. computer dialogue, programmed tuition)

- moving pictures to the centre, individually, (tuition television,
  remote diagnosis).

The first three types only need separate narrowband transmission
channels for each subscriber or only a few intervals in a broadband

time multiplex system, as the amount of information to be transmitted from the subscriber to the centre is relatively small. We may assume that there will be no bottlenecks in capacity, even if many subscribers should use a return channel at the same time.

However, the situation is quite different with the transmission of moving pictures from the subscriber to the centre. In the trial cable television facilities of the Deutsche Bundespost, it is possible, without laying any new cables, to have approximately four television channels in the return direction. Even if we assume that several cables of the highest network level, i.e. of the A level, are connected to the head station, there are return television channels available to all subscribers supplied by one head station in the order of 1 to 1000. We shall presumably have to expect in every case that a wide range of return channel services which also require the transmission of moving pictures in the return direction, will call for the laying of at least a second broadband cable.

We will reach similar conclusions if we look at the forward channel which is required for the various return channel services. As long as only comments on generally transmitted TV programmes are retrieved via the return channel or as long as it is used to determine the right of a particular subscriber to engage one or more special channels, for instance in the case of pay television, there are no particular problems involved in providing the forward channels. Here, the character of the cable television facility as a television distribution system remains basically unaffected and distribution channel capacity adequate.

Nor in the case of return channel services where texts or still frame pictures are transmitted individually to the subscriber, do there seem to be any serious problems, as we are dealing here with relatively small quantities of information which must be transmitted each time to the individual subscriber.

With the forward channels, too, the difficulties arise when moving pictures are to be transmitted individually to the subscriber. If we suppose that 10 per cent of all subscribers want to make use of such a service at the same time, then it would be necessary in a cable television network with 10,000 subscribers to have 1,000 broadband channels available. In my opinion, this considerably exceeds the range and scope of cable television networks, so that, when including individual broadband applications in cable television pilot projects, we can really only speak of experimental special cases.

However, let me take a step back in my observations. I said that the implementation of narrowband return channel services in cable television networks did not seem to present a major problem from the technical point of view. Nevertheless, there is still a "but", namely, the economic question. Narrowband return channel services are also possible in the telephone network, and are maybe even cheaper than in cable television networks. For, on the one hand, the splitting-up of broadband transmission channels into a number of narrowband channels is not without costs, and on the other, the telephone network already exists, as opposed to the cable television network.

C Closing remarks

I do not want to go into this any further, for we cannot yet say what results a cost comparison would produce. First of all, we must wait for the pilot projects to give us a clearer indication of the public's interest in return channel services. That is why we are very keen on speeding execution of the projects.

Kabelfernseh-Pilotprojekte in der Bundesrepublik Deutschland

Die vorliegende Arbeit gibt aus der Sicht der Deutschen Bundespost eine Darstellung des Standes des Kabelfernsehens in der Bundesrepublik Deutschland und zeigt, welche Rolle die Kabelfernseh-Pilotprojekte dabei spielen sollen.

Für die Deutsche Bundespost gibt es in diesem Zusammenhang zwei wichtige Fragestellungen:

Die erste Frage gilt dem zu erwartenden Publikumsinteresse am Kabelfernsehen, die zweite Frage der optimalen Gestaltung des technischen Systems, für das die Deutsche Bundespost die Verantwortung trägt, und zwar unabhängig davon, ob sie später auch das Netz selber errichtet und betreibt.

Die Deutsche Bundespost hat bisher drei sogenannte Kabelfernseh-Versuchsnetze errichtet, und zwar in Nürnberg, Hamburg und Düsseldorf. Diese Anlagen dienen der Wiederversorgung von Stadtgebieten, die durch Hochhäuser abgeschattet sind. Da jedoch nur die üblicherweise drahtlos empfangbaren Programme verteilt werden, handelt es sich im strengen Sinne nicht um Kabelfernsehanlagen, sondern um sehr große Gemeinschaftsantennenanlagen, die der Erprobung der technischen Bedingungen dienen sollen.

Da aus technischer Sicht für die Verteilung von Fernsehprogrammen kaum noch Probleme bestehen, richtet die Deutsche Bundespost ihr Hauptinteresse auf den Rückkanal und die mit seiner Hilfe realisierbaren Rückkanaldienste.

The Two-Way CATV Laboratory Project of the Heinrich-Hertz-Institute Berlin

Karl H. Vöge
Berlin, Federal Republic of Germany

Introduction

If asked what motivs make us run a major laboratory project in the field of broadband communication one may answer in the words of the KtK's Telecommunication Report (p. 167, English version)

> "On account of the little pronounced demand there is no reason to ask whether economic or social arguments speak against treating the development of the broadband distribution system as a matter of urgency. On the contrary, it should be examined whether the establishment of a nation-wide broadband cable system should be promoted, despite the unclear demand situation, for more important reasons affecting the public interest."

Accordingly, to enable more and even better telecommunication services for the public is the main development goal of the HHI two-way cable television (CATV) project, the present state of which I am pleased to report here.

Task Breakdown

Like all telecommunications systems a two-way CATV system, as depicted in fig. 1, consists of

- a transmitter, the central facility
- a transmission network, the coaxial cable network and
- one or more receivers, the subscriber home terminals.

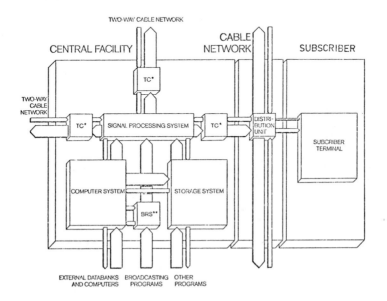

Fig. 1. Two-Way CATV System

In modern master antenna television (MATV) systems the reception
aerials for television and radio, together with the signal processors
for feeding the programmes into the network, constitute the central
facility. The cable network extends over active network levels (i.e.
ones equipped with amplifiers) and passive ones; as a rule, it is in
the form of a tree and normally distributes a maximum of six programmes
received over the air in a single cable.

At the end of the network the whole television signal, including radio,
is available to all subscribers, so that they can select the desired
programme on the television or radio set in their home. The subscriber
selects the programme, but direct influencing of the programme beyond
the range of what is offered is not possible.

With two-way CATV the aim is to extend these possibilities by enabling,
in addition,

- firstly information to be sent via the cable network from the
  subscriber to the central facility and

- secondly, individual information and programmes selected by the
  subscriber to be transmitted from the central facility to just
  a single subscriber.

For these operations to be performed additions must be made to the
technical installations of the central facility, the cable network
and the subscriber terminal.

The difficulty of defining the necessary technical additions lies in
the fact that the additional components are closely dependent on the
two-way contents desired or offered and on the actual use made of these
contents. Since, however, this question of use by subscribers is itself
still an important subject of investigation in the field experiments
planned in West Germany and at present can only be estimated approximately, the only possible basis for technical design of a future
dialogue system is a modular concept, i.e. one capable of adaptation.

For this reason, the Heinrich-Hertz-Institut für Nachrichtentechnik
Berlin, on request of the Federal Ministry of Research and Technology,
commenced work a year ago on laboratory construction of a two-way
CATV system of modular structure.

The purpose of this project in general is the development and demonstration of an interactive CATV system for 10,000 subscribers who
have access to a sufficiently complex menu of typical future services.
Since such public systems have not been realized yet it is necessary
to generate (fig. 2):

- Typical two-way services that take into account the present
  evolution of smallband and broadband services,

- A highly flexible CATV system concept containing interactive
  components within the central facility, the cable network and
  the subscriber terminals, and based on this

- A laboratory experimental CATV system as a guiding model for the
  technical design, for cost optimisation and for integration and
  experimentation of CATV services.

Only on this basis the required assistance for CATV field projects
is possible.

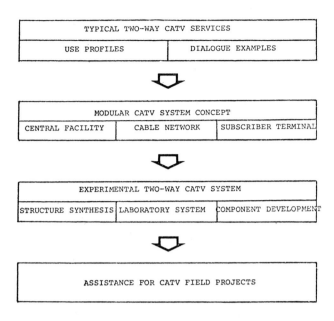

Fig. 2. CATV Project Requirements

As a result of more detailed analysis of these aims, the tasks involved were divided into five important groups:

1. Ascertaining typical two-way services

    - Determining dialogue algorithms
    - Specifying the requirements made on the technical system

2. Constructing a modular hardware model system

    - Experimental testing of the services
    - Theoretical description of the complete system

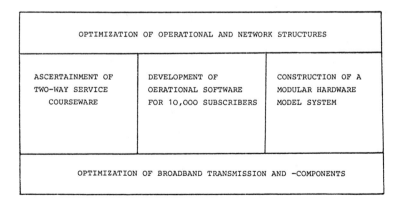

Fig. 3. CATV Project Tasks

3. Developing operational software for a 10.000-subscriber system
   - Theoretical description of the whole operation
   - Specification of a new operational system
4. Ascertaining optimal operational and network structures
   - Simulation of operational conditions
   - Investigation of optimal network topologies
5. Optimizing analogue broadband equipment
   - Technical optimizing of existing cable networks
   - Investigation of new transmission components.

With respect to HHI manpower each of these five fundamental tasks requires approximately the same amount of effort. This means about 200 man-months per task during the project's running time until the end of 1979. The 10-million DM project budget, however, breaks down into about 8,5 million DM for the three courseware, software and hardware tasks and about 1,5 million DM for the two optimization tasks.

Looking at some of the first working results in more detail, it should also become clear that technical development and economic optimization have to be extended to all system's components. That is to say that more traditional cost factors affecting individual communication systems, such as the cable network, play a lesser role here than the investment costs of the subscriber's equipment or the central facility. In addition there are new operational cost factors from the mass communications systems, such as, for instance, the provision and updating of the information.

Typical Two-Way Services

The great variety of conceivable services for cable television can be divided roughly into the following four categories (fig. 4).

CATEGORY	EXAMPLES
I Unlimited Distribution	National television, Local television, Open Channel, Sound radio
II Limited Distribution	Videotext, Pay television, Programme-recording, Electronic newspaper
III Dialogue	Information, Education, Agency services, Entertainment
IV Collection by the Central Facility	Remote control

Fig. 4. Four Categories of CATV Services

For two-way CATV category III, that of dialogue forms, is of particular interest. At the present stage of the project, in cooperation with a number of national and international experts and institutions, typical contents and structures of use are being worked out in classes and fields of application (fig. 5).

Class	Subclass/Examples	Stiftung Warentest	Dr.Schelske	ZBZ, Berlin	Prof.Haefner	Dr.Anders	Prof.Issing	Prof.Bunderson	HHI, Berlin	Use Profiles	Typical Examples	Ideal Service Requirements
Information	transport/ local, long-dist., town plan				x				x			
	culture/ cinema, theatre, sport medicine, authorities, aids								x			
	social services/ medicine, authorities, aids		x	x								
	market/ consumer information, tests	x										
Education	CMI/ tele-education				x	x			x			
	CCI/ tele-education				x	x	x	x				
	CAI/ tele-education man-machine (-man)						x					
Agency serv.	reservations/ cinema, doctors, authorities											
	purchases/ tele purchase of goods, services											
	bank services/ accounts											
	official services/ finances, social, labour		x	x								
Entertainment	games/ man-machine								x			
	games/ man-man											
Programming	Tutor							x				
	CA								x			

Fig. 5.   CATV Dialogue Services and HHI Working Contracts

Each group of users was given what was in principle the same task in their particular field of application:

1. Ascertain use data, e.g. extent of use access and reaction times, various dialogue possibilities.

2. Prepare typical examples of use in the form of detailed flow diagrams of the course of the dialogue that enable the examples to be taken over directly for the HHI laboratory system.

3. Describe an ideal system from the point of view of the user, regardless of technical factors.

From the first results of this work it is possible to estimate, for instance, the following dialogue requirements for a two-way cable television system with 10,000 subscribers connected (fig. 6, 7).

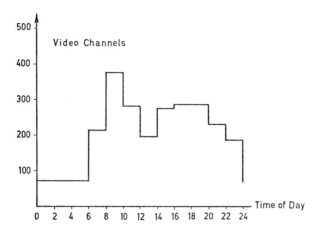

Fig. 6. Demand for Video Channels

Average daily time of use for each subscriber set	app. 1.2 hrs.
Time of most intensive use	app. 6-8 p.m.
Maximum no. of simultaneous dialogues	app. 800 subscribers
Average data flow downstream	app. 35 kbit/s
Maximum no. of audio channels downstream	app. 80 channels
Maximum no. of video channels downstream	app. 375 channels
Maximum no. of video channels upstream	app. 30 channels

Fig. 7. User Profile Figures

By means of such data a technical two-way system can now be specified in detail.

Constructing a Modular Hardware System

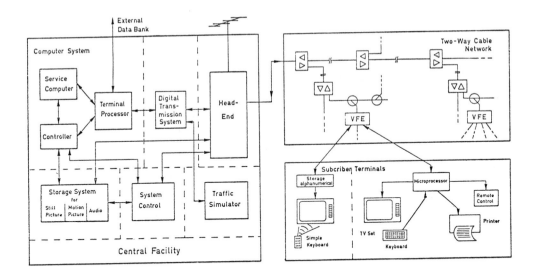

Fig. 8.   CATV Project Flow Diagram

The desired modular structure of this system (fig. 8) means that the necessary technical two-way supplementary components for present MATV technology fall into two groups, according to whether

- Immediate account has to be taken, as regards equipment and installations, of the full technical capacity expected to be used ultimately, or whether

- Technical extension is largely dependent on the contents desired and the number of subscribers.

The first group includes in particular:

- The coacial cable network from the central facility to the subscriber, which on account of the cost-intensive excavation work involved should be planned from the start to cope with the maximum demands expected later.

For future network-planning this means (fig. 9):

1. Dividing the 10.000 cable television subscribers into groups of about 1.000 subscribers each in a sub-area (HUB)

2. Connecting each of these network sub-areas via a separate cable with the central facility for transmitting the individually desired programmes of each sub-area (in the so-called adjacent neighbour channel position)

3. Retaining transmission of the present general distribution programmes in the present arrangement of channels on a second cable

4. Separating this second cable so that in the lower frequency range a smallband return channel service is possible from each subscriber to the central facility. In addition, in the lower frequency range a few (2-3) video channels should be capable of being transmitted from the subscribers in the direction of the central facility

5. Establishing star-form distribution of the individually desired programmes to the subscriber at the end of the tree network.

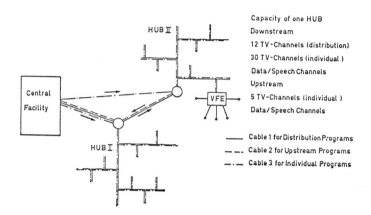

Fig. 9.   Schematic Representation of the Cable Network

In comparison with the KtK version with back conversion of the special channels at the television set, this means, according to our calculations, additional investment costs of about 25 % for the preparatory installations and about 240 % for a fully implemented two-way cable network.

The second group i.e. equipment which can be extended according to demand, through the addition of modules, includes in particular:

- The subscriber terminal and
- The central facility.

In addition to twelve television channels, time and programme indicators and channel tuning, commercial television sets in the near future will offer the facility of selecting Videotext. The necessary additional components for the subscriber terminals required for interactive television can be based on this.

An initial stage of television extension would thus be:

- A smallband data channel between each subscriber and the central facility and
- The possibility for the subscriber of receiving individually smallband and broadband programmes ordered from the central facility.

This means that every subscriber must have for his television receiver, in addition to some supplementary electronic units:

- A simple keyboard with a few symbols or figures, for instance in the form of an extension of the remote control unit in use at present,

- A memory for a limited amount of alphanumerical information in the television set (a refresh memory for an extended Videotext picture) and

- If desired, a simple printer connected to the set to store figures or short pieces of information.

For interaction the subscriber has available, for instance, an extended form of remote control unit, as in fig. 10.

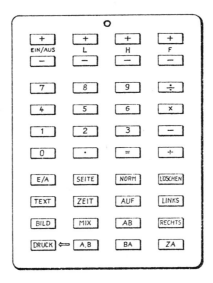

Fig. 10.   Remote Control Unit

The keys are arranged in three groups, according to their function:

- The top two rows of keys contain the customary television controls for switching on the set and regulating the volume, brightness and colour.

- The next four rows of keys correspond to those customary on basic pocket calculators. The figure keys can be used either for selecting programmes or for the input of figures.

- The remaining keys are function keys. With them the subscriber can control a cursor on the screen and, for instance, select from a proffered menu. He can control his printer with them, select Videotext pages, superimpose television picture and text, switch on a remote control unit, etc.

With these accessories nearly all forms of dialogue listed above can be carried out - although in some cases the quality and scope are restricted - in particular:

- Simple information retrieval systems (menu technique)
- System-controlled teaching programmes, table calculator functions
- Simple agency services (reservations, appointments, voting) and
- Games.

More complex forms of dialogue, such as learner-controlled learning systems or tele-purchasing with bank account debiting, require very much more sophisticated subscriber equipment, according to the particular demands. However, at this more sophisticated stage of extension the reservations that had to be made before concerning quality and flexibility no longer apply.

It is important that to offer these more complex forms of dialogue the principle of control and storage of contents by the central facility alone is given up in favour of the subscriber taking over these functions to an increasing extent in his terminal.

Technically, this means that the subscriber terminal must be provided with more "intelligence" and with greater storage capacity (for still- and motion pictures, possibly in colour, for audio and for programme control).

In terms of equipment this means that a television set, a radio, a more extensive alphanumerical keyboard, a printer for the DIN A 4 - format, a full video refresh memory, and also a camera or videorecorder, according to particular requirements, can be connected to an intelligent unit in the subscriber's home.

A final example of the modular system concept is the flexible layout of the computer system in the central facility. Fig. 11 shows various configurations for the three most essential parts of the computer system

- The service computer (SC), that is the dialogue partner to the subscriber and therefore mainly has to provide and manage the digital services
- The terminal processor (TP) that has to handle the i/o-procedures to and from the cable network which includes the address management of the subscribers and
- The controller (C), that controlls the central storage units for video and audio according to the subscribers demand.

Depending in increasing requirements the terminal processor may be bypassed or the service computer and the controller may be parallel organised.

It's important to state that two-way CATV systems can provide computer capacity to the home. This trend has been started i.e. by the development of intelligent stand alone terminals ("computer for the home") and by commercial courseware now offered to private use.

One of the main problems in the central computer system development
is

- The handling of very large numbers of simultaneously active subscribers

- The high quality required, i.e. reaction time ≤ 3 s, provision time ≤ 90 s and

- The high adaption level neccessary to serve the public spectrum of interest.

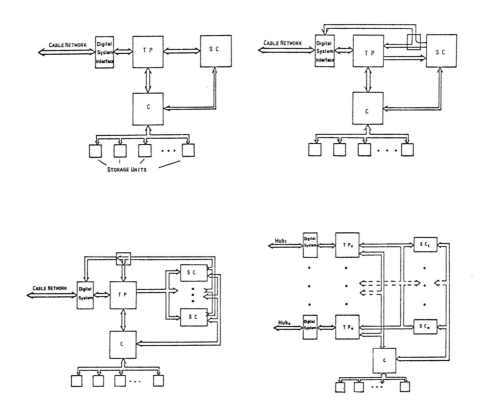

Fig. 11. Computer System Layout

Summary

The possibility not only of receiving visual and text information
but

- Of being able to transmit it oneself on "television" and
- Of having access to computer and databank capacity from
  the home

is, for individuals, the essentially new aspect of two-way CATV.

However, the new dialogue forms require more intensive laboratory
studies mostly with respect to operational software organisation,
to attractive service menus, to private data protection, to economics
of terminals and - may be sooner as aspected - to new transmission
media.

These will be decisive factors in determining acceptance as a means
of future mass communication.

Das Zweiweg-Kabelfernseh-Laborprojekt des Heinrich-Hertz-Instituts Berlin

Die Möglichkeit, nicht nur Bild- und Textinformation empfangen zu können,
sondern
- selbst Informationen per "Fernsehen" übertragen zu können und
- Zugriff zu EDV-Anlagen und Datenbanken zu haben

ist für den Benutzer der wesentliche neue Aspekt des Zweiweg-Kabelfernsehens.

Allerdings erfordern diese neuen Formen des Dialogverkehrs intensivere Labor-
untersuchungen im Hinblick auf die Organisation der betrieblichen Software, die
Bereitstellung einer Palette attraktiver Dienste, den Datenschutz im privaten Bereich, die Wirtschaftlichkeit von Teilnehmergeräten und - vielleicht früher als
vorauszusehen - den Einsatz neuer Übertragungsmedien.

Diese Faktoren werden darüber entscheiden, ob das Kabelfernsehen als künftiges
Massenmedium im Kommunikationsbereich zum Einsatz kommen wird.

Die Arbeit gibt eine Übersicht über das vom Heinrich-Hertz-Institut in Auftrag
genommene, modular konzipierte Laborprojekt, in dem typische Zweiweg-Tele-
kommunikationsformen für ein interaktives, 10 000 Teilnehmer umfassendes
Pilotprojekt vorgeführt und erprobt werden sollen.

Legal and Political Aspects of CATV Pilot Projects

Waldemar Schreckenberger
Mainz, Federal Republic of Germany

Technical development has created the pre-requisites for new systems in tele-communications. In a manner comparable to that of the transition to wireless broadcasting, a trend towards wire-borne information techniques is emerging. The new possibilities in the transmission and distribution of communication contents are not yet individually foreseeable. The products available from the electronics industry are contrasted up to now with a very limited preparedness to employ the new technical possibilities. Technical progress is experienced under the traditional pattern rather as a factor which could disturb the existing balance of interests and habits one has become accustomed to. In addition to this comes the fact that the introduction of new forms of communications is in part bound up with considerable financial expense. The technical impulse has not yet been followed by a corresponding economic or social drive. The political system also reacts in general with coutious reticence. The discussion about the intro - duction of new forms of communications is encountering a social mood which reacts to technical progress with scepsis and in part with a rebuff.

Information techniques are considered a particularly sensitive instrument for the dynamic forces of social change. The fear is quite widespread that the social structure could be influenced to its disadvantage by new forms of communications. Progress in media techniques undoubtedly entails risks, but also the opportunity of new possibilities for human self-realisation.

It was thus a correct decision by the commission appointed by the Federal Government (KtK) for the clarification of questions still open on subscribers' requirements and needs, acceptance and attitudes of use, to recommend that the CATV pilot projects be carried out. Even if one should not over-emphasize the chances of success of such trials, yet they appear imperative in order to make the social effects of technical developments a little more transparent. This is all the more true as the realisation of new technical possibilities can lead to a considerable change in individual and collective, social and private communications.

The carrying out of such model trials causes considerable legal and media-political problems.

As the responsibility for the whole matter is divided between the Federal and state governments, the question which first arises is that on legislative and administrative competence. One is very quickly confronted by a basic dilemma. This results from the fact that differing legal areas come into close contact, indeed partially encroach on one another.

On the one hand there is the Federal government's responsibility for posts and telecommunications, on the other the responsibility of the states for the organisation and utilisation area of broadcasting. If it is possible for individual new telecommunications services to be allocated to one or the other material, there are a series of kinds of utilisation of the networks which lead to quarrels between the Federal and state governments. The Federal government, for example, takes responsibility for the "screen text" because it entails a special recall by the subscriber by means of a telephone. On the other hand the states argue that the special recall can not be a criterion in those cases in which several people can switch in simultaneously into a presentation which is in progress. Can the "screen text" thus be a telephone service on one occasion and a radio one on another, depending on the technical construction?

The allocation of services to the fields of competence of the Federal government and states becomes more and more complicated if other objects of legislation especially mentioned in our constitution are touched upon. This is particularly true of the law on the press. Thus videotexts, which are made visible on the screen by electronic transmission, are claimed by the press. It regards the electronic transmission merely as a particular form of distribution of its products. The opposing argument is that everything which is made visible on the screen is to be allocated to broadcasting. It is completely immaterial whether it is here a question of fixed or moving pictures. The decisive factor is the distribution of information to the general public.

Other legislative responsibilities of the Federal government can also collide with the responsibility for broadcasting. Thus for example, questions of the right of the economy, problems of the prevention of abuse of economic positions of power and aspects of copyright law can gain importance.

Finally one should point to the fact that individual new techniques in literature
can be allocated neither to broadcasting nor to the press nor to any other written
legislative responsibilities. One should mention here in particular various
recall, warning and surveillance services. Such forms of communication are
in the regulatory competence of the states because the Basic Law -- the Constitution of our Republic -- assigns the exercise of all state jurisdiction to the
states so long as it does not expressly mention something else.

In the argument about the legislative competence, the struggle for the controlling
influence on the future shaping of our media organisation is given contours. It is
the term broadcasting which in the last resort determines the relationship of
tension. And here I must emphasise that the term broadcasting is defined by
the constitution. It is not subject to opportunistic considerations. It is immaterial
for the term broadcasting whether the information content is transmitted wire-
lessly or by means of a cable. I do not fail to realise that the term broadcasting
must be sonsonant with technical development if it is not to stand in the way of
modern transmission of information.

A further problem arises from the present structure of our broadcasting system
which in contrast to the private press is organized on a public law basis. The
dualism of public law broadcasting and a private press is defended by interested
parties as being a natural state of affairs. They fear that new forms of communication could open the fixation of this dualism to doubt. Others see in new forms of
communication an opportunity to overcome the existing broadcasting organization.
Anyone who burdens the pilot projects with all these problems is hardly likely to
arrive at a suitable decision. It can neither be a question of having doubts on
principle about the existing structure in the media field, nor should it be forbidden
to try alternative forms of organization.

For the central question is: How can the new foreseeable communication techniques
be used sensibly? Public law or private law structures are not authoritative
criteria here from the point of view of the constituion. Much more decisive is
rather the guarantee of a public communication process in which a variety of
opinions is expressed. The main aim should be to expand and improve the
citizens' chances of obtaining plurality of opinions with the aid of new technologies.
An automatic result of this is that in pilot projects alternative and competing
organizational forms should be tried out with the participation of as many
institutions and social groups as possible.

Only in this way can sufficient knowledge be gained on its resonance among subscribers. That is the pre-requisite for objectively correct decisions on social-politically desirable organizational structures.

Similar considerations apply for the utilisation contents. They must be as many-sided as possible. Only in this way can the acceptance be tested. Apart from the local programmes of the public law broadcasting stations which can not be received without wire, which should also be offered at a later time as repeat broadcasts and in a particular combination, it is the new types of services which come particularly into consideration. One should think here of programmes with local relevance, of video and screen texts, of programme contributions by third parties in an open channel available to all, of cable text recalls with reply channels and individual texts. Sectional programmes and pay-TV should not be excluded. If the objection is raised against such programmes that they would lead to an endangering of the programm, organizational and financial bases of the existing broadcasting systems, then I must object that it is precisely the purpose of pilot projects to give information on such questions. What is socially desirable can not be decided in advance. This would lead to the citizens' being deprived of a decision on new forms of communications. Such a caesura would, in the last resort, mean patronage by the state.

Model trials have particular importance for the press. As its ability to function is guaranteed by the constitution, pilot projects must be so arranged that reliable statements on the effects of new forms of telecommunications on newspapers and periodicals are possible. This is true to a particular degree for the local and regional press. In particular a study must be made on whether an increase of radio and television programmes with local, regional and supraregional relevance with advertising and commercial spots would so shatter the economic basis of the daily newspaper that the monopolisation of the press further continues. But a study must also be made whether on the other hand many-sided services by the broadcasting services do not help against monopolistic tendencies in the press.

The financing of pilot projects presents special problems. In view of the risks involved in the test phase and with regard for the fact that only a very restricted number of subscribers can be included in pilot projects, nobody has been prepared up to now to take greater financial risks. Progress was always coupled with risks. It is not correct to require of the state that it show the interested parties their opportunities at no risk to themselves.

Admittedly the financial commit ment for private organizers must be kept within limits. Technology has after all caused the state to examine how the technical possibilities can be used to the citizens' best advantage.

Only about 10. 000 households should be covered in a pilot project. In order to test the acceptance, the utilisation costs for the subscribers must thus be kept as low as they would probably be with appropriate spread under cost-covering conditions of permanent operation. On should thus strive for a financial compromise. The basic rule must be that all projects should have the same chance and are thus suitable for comparison. This means that even private sponsors must also receive public grants. This is true both for the network and also for the utilisation area. The Federal government is responsible for the co-financing of the network to which the final apparatuses or additional equipment belong. In the separation which is demanded of the network area from the utilisation area, there are no objections to the Federal government's making finance available. The utilisation area which comes within the states' competence can probably only be financed in a mixed form. As sources of finance, apart from fees paid by the subscribers, free contributions by the organizers, a continuous funding from the resources from broadcasting licence fees and grants from the states only come into consideration on condition that adequate measures are taken against state influence on the programmes.

The states have not yet arrived at a common answer to all points in the questions raised. The decision on carrying out pilot projects has yet to be made. The linking together of media political and economic risks of considerable magnitude has done nothing towards speeding up the decision. Even the comparison with experience gained abroad does not seem very encouraging to many. It is thus understandable that individual interested parties try to anticipate parts of the trials, such as the screen neewspaper. However, by doing this they make the process of reaching a decision more difficult through which the way would be opened for a controlled expansion of our communications system. Pilot projects are only of value if they include all those new forms of telecommunications which come in for consideration. For only in-this way can they be a suitable basis for the further development of our media organization.

Rechtliche und politische Aspekte der Kabelfernseh-Pilotprojekte

Auf dem Gebiet der Telekommunikation ist der technischen Entwicklung noch kein entsprechender ökonomischer und politischer Impuls gefolgt. Dieser Fortschritt bedeutet Risiko, aber auch neue Chance der Kommunikation. Der Verfasser setzt sich daher für die Durchführung von Pilotprojekten ein, in denen alternative Organisationsformen und Nutzungsinhalte erprobt werden. Er spricht sich dafür aus, die Durchführung solcher Versuche nicht mit den Grundsatzfragen der Rundfunkpolitik, insbesondere der Frage der öffentlich-rechtlichen oder privatrechtlichen Struktur des Rundfunks, zu belasten. Er wendet sich an mögliche private Programmträger, mehr Unternehmermut als bisher zu zeigen. Es sei allerdings Sorge dafür zu tragen, daß das wirtschaftliche Risiko auch für einen privaten Träger begrenzbar bleibt. Daher fordert der Verfasser, daß nicht nur öffentlich-rechtliche Träger, sondern auch private Veranstalter öffentliche Zuschüsse erhalten. Pilotprojekte sind nur sinnvoll, wenn sie alle in Betracht kommenden neuen Formen der Telekommunikation erfassen. Der Verfasser weist dabei auf die Probleme hin, die sich aus der Aufteilung der Gesetzgebungs- und Verwaltungszuständigkeiten zwischen Bund und Ländern ergeben.

List of Contributors and Chairmen

T. F. Baldwin, Prof. Dr.

Michigan State University
Department of Telecommunication

East Lansing / Michigan 48823 / USA

F. Baur, Dr. -Ing.

Siemens AG
Unternehmensbereich Bauelemente
Balanstr. 73

8000 München 80

Federal Republic of Germany

Mrs. J. Bazemore, Dr.

Rand Washington Office
2100 M Street, N. W.

Washington, D. C. 20037 / USA

L. Bonavoglia, Prof. Dr. -Ing.

S I P - Direzione Generale
Via Flaminia 189

I - 00 196 Roma / Italy

J. L. Bordewijk, Prof. Dr.

Technische Hogeschool Delft
Afdeling der Elektrotechniek
Mekelweg 4

Delft 8 / Netherlands

Ch. N. Brownstein, Prof. Dr.

Program Manager
Advanced Productivity Research
and Technology Division
National Science Foundation
1800 G Street N. W.

Washington, D. C. 20550 / USA

E. J. Callahan Jr.

Director of Research
American Television +
Communications Corporation
360 South Monroe Street

Denver, Colorado 80 209 / USA

W. Kaiser, Prof. Dr. -Ing.

Institut für Nachrichtenübertragung
der Universität Stuttgart
Breitscheidstr. 2

7000 Stuttgart 1

Federal Republic of Germany

J. Kanzow, Dipl. -Ing.	Ministerialrat Bundesministerium für das Post- und Fernmeldewesen Postfach 80 01 5300 Bonn 1 Federal Republic of Germany
M. Kawahata, Dr.	Managing Director V I S D A Sanko Bldg. 4-10-5 Ginza Chuo-ku, Tokyo 104/Japan
D. Kimbel, Dr.	O E C D 2, rue André Pascal 75 775 Paris Cedex 16 / France
W. A. Lucas	Rand Washington Office 2100 M Street, N. W. Washington, D. C. 20037 / USA
H. Marko, Prof. Dr. -Ing.	Institut für Nachrichtentechnik der Technischen Universität München Arcisstr. 21 8000 München 2 Federal Republic of Germany
W. F. Mason	Technical Director Systems Development The Mitre Corp. Research Park McLean, Virginia 22101 / USA
G. Moreau	Technical Adviser c/o E B E S Franklin Roosevelt Laan 1 B-9000 Gent / Belgium
M. L. Moss, Prof.	Associate Professor of Planning and Public Administration New York University 4 Washington Square North New York, N. Y. 10003 / USA
H. Nakajima	Staff Engineer Engineering Bureau N T T 1-6, Uchisaiwai-Cho 1-Chome Chiyoda-Ku Tokyo 100 / Japan

T. Namekawa, Prof. Dr.

Department of Communication Eng.
Faculty of Engineering
Osaka University
Yamadakami, Suita

Osaka / Japan

W. Neu, Dr.

Technisches Zentrum
P T T , V706

CH-3000 Bern 29 / Switzerland

K. C. Quinton

Director of Research
Rediffusion Engineering Ltd.
187 Coombe Lane West

Kingston-upon-Thames
Surrey KT2 7DJ / Great Britain

J. Rottgardt, Dr.

Standard Elektrik Lorenz AG
Hellmuth-Hirth-Str. 42

7000 Stuttgart 40

Federal Republic of Germany

W. Schreckenberger, Prof. Dr.

Staatssekretär
Staatskanzlei des Landes
Rheinland-Pfalz
Rheinstr. 113

6500 Mainz

Federal Republic of Germany

C. Sechet

Centre Commun d' Etudes
de Télévision et Télécommunications
B. P. 1266

F-35013 Rennes / France

I. Switzer

P. Eng., Consulting Engineer
Switzer Engineering Services Ltd.
5840 Indian Line

Mississauga, Ontario L4V 1G2

Canada

K. H. Vöge, Dr.

Heinrich-Hertz-Institut
Einsteinufer 37

1000 Berlin 10

Federal Republic of Germany

Mrs. S. Walters

President
Scarboro Cable TV/FM
Unit 33 705 Progress Avenue

Scarborough, Ontario / Canada

H. Weber, Dr. -Ing.

AEG-Telefunken
Nachrichten- u. Verkehrstechnik

7150 Backnang

Federal Republic of Germany

E. Witte, Prof. Dr.

Institut für Organisation der
Universität München
Amalienstr. 73 b

8000 München 40

Federal Republic of Germany

J. v. Wrangel, Dipl. -Ing.

AEG-Telefunken
Gerberstr. 33

7150 Backnang

Federal Republic   of Germany

K. Yamaguchi

Representative Director
Seikatsu Joho Shisutemu
Kaihatsu Honbu
Gurinado Nagayama
4, Nagayama 1-Chome

Tama-shi, Tokyo / Japan

S. Yoshida, Dr.

Technical Counsellor
Seikatsu Joho Shisutemu
Kaihatsu Honbu
Gurinado Nagayama
4, Nagayama 1-Chome

Tama-shi, Tokyo / Japan

Nachrichten-technik

Herausgeber: H. Marko

Band 1
H. Marko, Technische Universität München

Methoden der Systemtheorie

Die Spektraltransformationen und
ihre Anwendungen

87 Abbildungen, 11 Tabellen. Etwa 200 Seiten. 1977
ISBN 3-540-08106-2

Band 2
P. Hartl, Technische Universität Berlin

Fernwirktechnik der Raumfahrt

Telemetrie, Telekommando, Bahnvermessung

104 Abbildungen. XIII, 208 Seiten. 1977
ISBN 3-540-08172-0

Band 3
E. Lüder, Universität Stuttgart

Bau hybrider Mikroschaltungen

Einführung in die Dünn- und Dickschichttechnologie

141 Abbildungen. Etwa 160 Seiten. 1977
ISBN 3-540-08289-1

Band 4
H. Kremer, Technische Hochschule Darmstadt

Numerische Berechnung linearer Netzwerke und Systeme

29 Abbildungen, 26 Tabellen. Etwa 200 Seiten. 1977
ISBN 3-540-08402-9

Springer-Verlag
Berlin
Heidelberg
New York

Taschenbuch
der Informatik

In drei Bänden. Unter Mitwirkung zahlreicher Fachleute

Herausgeber: K.W. Steinbuch, W. Weber
Redaktion: T. Heinemann
3., neubearbeitete Auflage des Taschenbuches der Nachrichtenverarbeitung (1. und 2. Auflage erschienen unter dem Titel „Taschenbuch der Nachrichtenverarbeitung")

Band 1
Grundlagen
der technischen Informatik

429 Abbildungen. XIII, 563 Seiten. 1974
ISBN 3-540-06240-8

Band 2
Struktur und Programmierung
von EDV-Systemen

362 Abbildungen. XIV, 672 Seiten. 1974
ISBN 3-540-06241-6

Band 3
Anwendung
und spezielle Systeme
der Nachrichtenverarbeitung

302 Abbildungen. XIII, 463 Seiten. 1974
ISBN 3-540-06242-4

Aus den Besprechungen:
„Auch die dritte Auflage des weithin bekannten Werks verfolgt das Ziel, ein Nachschlagewerk für alle unter dem Begriff der Informatik zusammengehaltenen Gebiete darzustellen und dabei hinsichtlich Elektronik, Systemanalytik, Datenorganisation usw. möglichst umfaßend zu sein. Das nunmehr dreibändige Werk ist in Klarheit der Erklärung, Ausführlichkeit der Darstellung, Abstimmung der Beiträge untereinander und Übersichtlichkeit als wahrhaft vortrefflich zu bezeichnen. Die große Zahl der an Datenverarbeitung Interessierten wird das dreibändige Taschenbuch sicherlich als Spitzenwerk einstufen."
Elin-Zeitschrift

K.W. Steinbuch

Automat und Mensch

Auf dem Weg zu einer kybernetischen Anathropologie
4., neubearbeitete Auflage
131 Abbildungen. VII, 266 Seiten. 1971
(Heidelberger Taschenbücher, Band 81)
ISBN 3-540-05154-6

Aus den Besprechungen:
„Der Untertitel dieses faszinierenden Buches lautet „Auf dem Weg zu einer kybernetischen Anthropologie". Tatsächlich läuft als roter Faden durch diese vorzügliche Einführung in das Wesen und die Möglichkeiten nachrichtenverarbeitender Systeme das Verständnis geistiger Vorgänge auf Grund von bekannten physikalischen Prinzipien. Dabei treten zu den materiellen und energetischen Zusammenhängen eine Reihe von Informationsrelationen. Das Buch verhilft somit zu einem tiefen Verständnis der menschlichen und maschinellen Verhaltensweise und stellt viele konventionelle Denkschemata in Frage. Als logischer Schluß ergibt sich, daß der Mensch sich nur in der absoluten Freiheit von mystischen Vorstellungen geistig weiterentwickeln kann. Das auch für den Nichtwissenschaftler verständlich geschriebene Buch ist reich und anschaulich illustriert."
Neue Züricher Zeitung

Springer-Verlag
Berlin
Heidelberg
New York